Agile eHealth Usability Evaluation

Irina Sinabell

Agile eHealth Usability Evaluation

Development of the ToUsE toolbox to foster the usability of eHealth systems

Irina Sinabell
Wien, Austria

Doctoral thesis UMIT TIROL, 2023

ISBN 978-3-658-44433-4 ISBN 978-3-658-44434-1 (eBook)
https://doi.org/10.1007/978-3-658-44434-1

This Springer Vieweg imprint is published by the registered company Springer Fachmedien Wiesbaden GmbH, part of Springer Nature.
The registered company address is: Abraham-Lincoln-Str. 46, 65189 Wiesbaden, Germany

Paper in this product is recyclable.

The most damaging phrase in the language is: It's always been done that way.

Grace Hopper

Foreword

The aim of patient empowerment, patient participation and patient co-creation is to significantly and positively impact the development of effective IT systems and platforms in health care. The ongoing development of these systems and platforms creates an increasingly diverse and complex health care IT environment, which makes it particularly important to ensure its user-friendliness, especially for its adoption by different groups of people as potential users. Who are we to decide what a particular group of people – citizens, patients, doctors or even more specialized individuals – need and do not need from an app, device, system or platform? This is in line with the concept of universal usability, where the aim is for health care IT to be tailored to the needs of different potential user groups.

To fully determine the impact, usability and benefits of health care IT, adapted evaluation methods are needed with regard to more agile development approaches including co-creation. This is especially true for usability evaluations, as the evaluation with prospective users must be deeply integrated into the development process. The needs of prospective users also entail different patterns of use, which manifest themselves in the digital divide between older and younger people. Care must be taken to ensure that health care IT is not only developed for young people with IT knowledge. Older, chronically ill or disabled persons with limited IT skills are an important target group as well. Useful, patient-centered eHealth interventions are therefore vital to reach this target group. In this book, patients and their health care needs are placed at the center of health care IT in order to improve health care systems and platforms. Another decisive factor is the suitability of an agile eHealth usability evaluation for the agile development process of patient-centered eHealth interventions. Unfortunately, the translation of projects into daily routine in health care settings often fails because usability was not evaluated with prospective users or because certain user groups were underrepresented. Improved usability is therefore important to counteract failures in health care. Safe and usable eHealth interventions are crucial in health care, as system failures can have serious consequences. The usability of health care IT is therefore an important concern, as usability issues with eHealth technologies can put patients at risk. Consequently, challenges that may arise during an agile eHealth usability evaluation with prospective users need to be countered and addressed.

It is important to investigate how agile usability evaluations can be implemented into agile development processes of patient-centered eHealth technologies. The aging of society and its specific needs and requirements require us to use evaluation methods that take these needs into account and address them. This book is devoted to the improvement of eHealth usability, leading to increased patient-centeredness of eHealth interventions in order to contribute to achieving universal usability in the information society. It provides an insight into 43 different evaluation methods and in a case study evaluates how these methods can be applied in the development of eHealth technologies for the elderly. The findings presented in this book encompass prioritized eHealth usability evaluation methods that are pertinent to patient-centered eHealth interventions, contributing to meaningful improvements in the usability of eHealth interventions and thus enabling accessible health care for the aging population. The book ends with a discussion of how the research findings will facilitate the use of agile eHealth usability evaluations as a step towards universal usability of eHealth interventions.

November 23, 2023

Björn Schreiweis
Professor for Medical Informatics
Kiel University and
University Hospital Schleswig-Holstein
Kiel, Deutschland

Preface

Empathy is essential in creating technology that truly serves the needs of all users.

Margaret Hamilton

Human-Centered Design stellt die humanisierte Entwicklung neuer Technologien in den Fokus. Deren Anwendung ist gerade in kostenintensiven Bereichen wie der Gesundheitsversorgung geboten. Dabei rückt die agile Softwareentwicklung in den Vordergrund, welche weniger den traditionellen Weg der Entwicklung verdrängen, sondern in diesem spezifischen Kontext der Gesundheitsversorgung den Umgang mit niedrigen Entwicklungsressourcen zu gestalten vermag, also in einem strukturellen Zusammenhang mit dem Gesamtspektrum der Gesundheitsversorgung zu betrachten ist.

Der Dualismus zwischen agil und traditionell ist somit nicht im Sinne von modern und konservativ zu sehen, vielmehr soll die Auswahl des Weges zuvorderst kontextabhängig erfolgen. Ein kostenintensiver Bereich wird im Gesundheitswesen auf den institutionellen Bereich, vor allem für die stationäre und ambulante Versorgung von Patientinnen und Patienten, verwendet. Der WHO zufolge ist die Lebenserwartung in den letzten Jahren in Österreich angestiegen. Dies mag auf verschiedene Faktoren, wie etwa den medizinischen Fortschritt, einen steigenden Lebensstandard oder bessere Ausbildung zurückzuführen sein. Ein Faktor dieser Veränderung liegt damit auch in der Entwicklung und Nutzung neuer Technologien im Gesundheitsbereich, die neue Behandlungsmethoden und Präventionsinstrumente hervorgebracht haben. Optimalerweise führt das

Zusammenwirken dieser Faktoren insgesamt gesehen zu einer „gesünderen" Gesellschaft, im Besonderen im ersten und zweiten Lebensdrittel.

Umgekehrt und trotz dieser Rahmenbedingungen bedeutet eine höhere Lebenserwartung im Leben eines Menschen tendenziell auch die Verlängerung jener Phasen, in denen er einer medizinischen Betreuung bedarf und angesichts sich steigernder Abhängigkeiten der Gedanke der Selbstbestimmung immer mehr an Bedeutung gewinnt. Dabei sind gesamtgesellschaftlich gesehen grundlegende Ressourcenfragen und auch die Gegebenheit in Erwägung zu ziehen, wie die „Aufgabenverteilung" in der medizinischen Versorgung zwischen den öffentlichen Institutionen und der Gesellschaft als solcher, also jeder/jedes Einzelnen, ausgestaltet ist. Eine gewisse Selbstversorgung, die heute freilich durch die genannten neuen Technologien tendenziell erleichtert wird, ist nicht nur wünschenswert, sondern sogar erforderlich. Patientenzentrierte Technologien fokussieren sich auf Patienten als zukünftige Benutzer und deren Gesundheitsbedürfnisse, um die Möglichkeit einer Selbstversorgung zu bieten. Die Handhabung neuer patientenzentrierter Technologien muss benutzerorientiert erfolgen, wenn es nicht zu nachteiligen Auswirkungen, etwa durch Fehlfunktionen oder die falsche Anwendung dieser Technologien, kommen soll.

Hier kommen sowohl das Thema als auch das Anliegen meiner vorliegenden Arbeit ins Spiel. Sie ist von meiner Motivation und dem Wunsch getragen, auf Basis umfassender empirischer Methoden ein evidenzbasiertes Bild einer agilen Usability Evaluation dieser Instrumente durch ältere Personen zu entwerfen und in der wissenschaftlichen Herangehensweise und im Besonderen für die Durchführung entsprechender Studien eine Toolbox zur agilen Usability Evaluation »ToUsE« zu entwickeln, auf der weitere Forschungen in diesem Gebiet aufbauen können. Neben dem wissenschaftlichen Ertrag rückt dabei auch der Aspekt in den Blick, dass es letztlich um den Dienst am Menschen geht, wenn wir uns mit einer agilen Usability Evaluation und deren wissenschaftlichen Durchdringung vor dem Hintergrund einer alternden Gesellschaft beschäftigen, die gemeinhin von einer zuweilen beschwerlichen oder beschwerlicher werdenden Lebenssituation geprägt ist. Die medizinischen Wissenschaften sind seit Hippokrates dem Dienst am Menschen verbunden, und das Gleiche gilt auch für die Erträge der Wissenschaft, die Modalitäten der medizinischen Versorgung der Menschen analysiert und gerade aufgrund ihres Schnittstellencharakters interdisziplinär angelegt werden muss.

Solcherart versteht sich meine Arbeit als im Dienst des einzelnen Menschen, aber auch der Gesellschaft an sich stehend. Wie in dem einführenden Zitat zum Ausdruck gebracht, ist die Motivation, den Bedürfnissen einer großen Menge an Nutzern gerecht zu werden, dem übergeordneten Ziel meiner Arbeit, nämlich dem Erreichen einer universellen Usability neuer Technologien, gedanklich zuzuordnen. Die Botschaft dieses Zitats hebt die Bedeutung von Empathie bei der Entwicklung neuer Technologien hervor, die auch als die Bereitschaft eines Forschers, unterschiedliche Anforderungen einer Gesellschaft bei der Entwicklung neuer Technologien zu berücksichtigen, verstanden werden kann. Das Thema meiner Arbeit ist daher zugleich aktuell wie zeitlos. Indes ist sie damit auch nicht mehr als eine Momentaufnahme der Forschung, die darauf angewiesen ist, dass auf ihrer Basis weitere Forschungen betrieben werden. Dazu ist es erforderlich, dass ihre Ergebnisse im Wege einer Veröffentlichung einer breiteren Diskussion zugänglich gemacht werden.

Wien
im November 2023

Irina Sinabell

Acknowledgements

I am grateful to a number of people who have encouraged me at all stages of the realization of this work.

First and foremost, I would like to acknowledge Prof. Elske Ammenwerth for her guidance, continuous discussions, and expertise during my doctoral program in health information systems. Our research discussions always offered me new perspectives, and she provided me with contacts abroad that led to enriching experiences at universities in other European countries and gave me valuable insights into relevant research in the field.

I also appreciate the support of Ass.-Prof. Linda Dusselje-Peute of the Academic Medical Center at the University of Amsterdam, who motivated me in my doctoral project and took time to discuss current trends in human–computer interaction.

I further would like to thank Prof. Christian Nøhr for enabling me to participate in a research seminar in Denmark, which gave me valuable insights into the work of SDU Health Informatics and Technology and the Maersk Mc-Kinney Moller Institute at the University of Southern Denmark.

My thanks go to all the participants who took part in the study that was necessary for realizing this work. I am especially thankful for the time and expertise of the human–computer interaction specialists who made it possible to conduct the iterative validation of easily applicable and useful methods to support faster eHealth usability evaluations. The valuable feedback from elderly participants facilitated the implementation of the explorative case study to develop recommendations for addressing challenges during eHealth usability evaluations. Finally, I would also like to thank MediPrime for sharing their newly developed web-based eHealth invention for implementing the eHealth usability evaluation.

About the Book

This book represents the revised summary paper to my doctoral thesis, titled "Innovation in health care: Agile usability evaluation for patient-centered eHealth inventions", which I defended in 2023. The topic is highly relevant from both a methodological and a practical point of view. As far as the methodological aspect is concerned, many – unfortunately often complex – methods are currently applied to accomplish a usability evaluation. Little consideration, however, has so far been given to how to carry out usability evaluations for eHealth interventions in an "agile" way.

Health information systems such as mHealth or pHealth interventions are frequently subject to rapid development cycles, which makes it crucial to address agility already in the early stages of software development, especially with regard to the expected usability. Usability evaluations should therefore be carried out continuously in order to constantly improve the usability of eHealth interventions. In the context of my present work, I refer to eHealth systems as eHealth interventions used with the aim of improving patient self-care or of supporting patient therapy. As part of this work, I examined whether and which usability evaluation methods are suitable for this area of application.

eHealth interventions are often intended for older or cognitively impaired people. Examining usability is therefore vital for the acceptance and thus for the usefulness of eHealth interventions. eHealth interventions should not be developed only for "people like me", i.e., young, tech-savvy people who know how to use a computer. An important future user group of eHealth interventions are older, chronically ill patients who may have limited IT skills. User-friendly eHealth interventions are therefore crucial to ensure that this prospective user group is reached. Agile usability evaluations can make a significant contribution in this regard.

The present work is based on two publications and one research report. These are made available as part of the book's electronic supplementary material (ESM).

Contents

List of Figures

List of Tables

1 Introduction

This book examines how an agile usability evaluation for patient-centered eHealth inventions can be realized to foster usability evaluations in health care that are easy to implement and rapid to perform. This section provides background information on evaluating the usability of eHealth inventions, taking into account the iterative approach of software development, which is characterized by short iterations designed to optimize a part of a software.

Traditional concepts of usability evaluation, such as user-based and expert-based usability evaluation methods, which are difficult to reconcile with the iterative approach to agile software development, are presented. This introduction focuses on the elderly as a prospective user group for eHealth inventions, the challenges involved, and the need to offer improved eHealth usability for this user group.

The introduction describes the interfaces between usability engineering, user experience, and human-computer interaction. It addresses the relation of human-computer interaction to universal usability for eHealth. Agility for eHealth and models of the software development life cycle are also discussed. The research gap is framed in relation to answering the research questions and addressing the research subject presented in the introduction.

Key terms used in this book are explained in Table I that can be found in the glossary and are shown in italics when first mentioned.

Supplementary Information The online version contains supplementary material available at https://doi.org/10.1007/978-3-658-44434-1_1.

I. Sinabell, *Agile eHealth Usability Evaluation*,
https://doi.org/10.1007/978-3-658-44434-1_1

1.1 Innovations in Health Care, the Aging Population, and Agile eHealth Usability Evaluation

Design for *elderly* users should be simple (Razak et al., 2013). Innovations in the field of *eHealth* are emerging very quickly (Broekhuis et al., 2019a), leading to an increase in the complexity of eHealth inventions. Current advances in information technology are improving the quality of life (Jakkaew & Hongthong, 2017), and overarching *information and communication technology*-based innovations can be used for social interaction (Schwaninger et al., 2022) or to maintain health (Spanakis et al., 2016). These opportunities are leading to increasing use of technology-based products and services in our daily lives, thus transforming our society into an information society (Leahy & Dolan, 2009). Recent years have witnessed the rapid development of information and communication technology (Oikonomou et al., 2009) concurrently with the aging of the population, leading to a digital divide between elderly and younger people in terms of utilizing information and communication technology (Gilbert et al., 2015). *eHealth systems* can help the elderly to live independently (Sülz et al., 2021). This rapid technological progress is leading not only to higher complexity, but also to an increased variety of eHealth inventions.

eHealth inventions contribute to improved health care (Blankenhagel, 2019), enabling affordable and accessible health care in rural regions (Chiarini et al., 2013). In the context of this work, eHealth interventions are defined as eHealth inventions that are used in a supportive, preventive, or therapeutic manner. eHealth interventions are regarded as tools or treatments that are typically behavior-based, operationalized, and transformed for delivery via the internet (Eysenbach, 2018). Some eHealth interventions target vulnerable *user groups* such as patients who suffer from diabetes (Mayberry et al., 2019), pregnant women (Fryer et al., 2020), or children (Knapp et al., 2011) who may belong to marginalized groups (Spiel et al., 2020). Within the scope of this book, marginalization refers to user groups that are disabled or elderly. Recent research using the United Kingdom as an example shows that digital health initiatives to combat COVID-19 exclude certain user groups, e.g., groups with lower social status, low education levels, or the elderly (Sounderajah et al., 2021). These populations have a low propensity to adopt eHealth systems, which, coupled with the aging of the population and the increase in chronic diseases, is leading to new challenges in health care (Góngora Alonso et al., 2019). Meeting these burdens on the health care system will require new health care solutions that can be addressed with the use of eHealth systems (Cunha et al., 2014). eHealth systems are used to inform, "prevent, diagnose, treat, or monitor health conditions" (Broekhuis et al., 2021); they are used for medical reasons (Arning & Ziefle, 2009) as well as to improve the mobility of the elderly. eHealth encompasses a variety of *patient-*

centered eHealth systems, such as electronic health records (Aguirre et al., 2019), patient information websites (Peute et al., 2015), clinical decision support systems (Bonis, 2019), and mobile health systems (*mHealth* systems) (Lazard et al., 2019). For instance, mHealth systems can be applied for personal health care to accomplish continuous monitoring of chronic diseases. Medical data, such as results from blood tests, is collected via mHealth systems and sent to the hospital for decentralized health care.

The eHealth systems designed for decentralized health care target not only vulnerable but also various prospective user groups. For instance, there are eHealth systems for children (Ramsey et al., 2020), adolescents (McCann et al., 2019), adults (Badawy et al., 2018), and the elderly (Villani et al., 2016). eHealth systems can also be used to provide perinatal care (van den Heuvel et al., 2018). The rationale for this diversity of eHealth systems is that the different morbidities of the various prospective user groups present unique challenges to health care services. As an example, eHealth systems have been developed for both children with mental illness (Iorfino et al., 2019) and elderly patients with chronic illness (Lum et al., 2017). Elderly patients usually suffer not only from a chronic disease, but also from multimorbidity (Pohl et al., 2016). In summary, it is obvious that many different types of eHealth systems have been developed, deployed, and studied in the field in recent years (Kip et al., 2022).

The Austrian health care system is well equipped and prepared for the current and steadily increasing use of eHealth systems in the future (Wernhart et al., 2019). Broadband internet, which enables a high data transfer rate, is available nationwide in Austria (Bundesministerium für Verkehr, Innovation und Technologie (BMVIT), 2019). The further expansion of the communications infrastructure, crucial for the use of information and communication technologies, serves as the basis for digitization in Austria and represents a key part of the country's Broadband Strategy 2030 (Bundesministerium für Verkehr, Innovation und Technologie (BMVIT), 2019). However, Austrians are not very familiar with the possibilities of these ample eHealth systems (Haluza et al., 2016). In other European countries, the development of innovative eHealth systems for younger people is on the rise (McCann et al., 2019) and eHealth systems for elderly users are becoming more common (Wildenbos et al., 2019b). Elderly people are generally in favor of utilizing health-related technologies (Arning & Ziefle, 2009) such as eHealth systems. They are willing to adopt new technologies (Jakkaew & Hongthong, 2017) but sometimes find it difficult to keep up with the rapid technological advances and to maintain the experience level of younger people. Ensuring the well-being and quality of life for the elderly has become a priority for modern society (Vitiello & Sebillo, 2018).

This illustrates the need to keep the design of eHealth systems simple in terms of their ease-of-use and improved *usability* for elderly users.

Due to the rapidly progressing development of eHealth systems and their associated rise in complexity, the usability of these systems is becoming an increasingly important topic (Gilbert et al., 2015). Usability is regarded as a quality feature concerning the use of software (Magües et al., 2016) that includes aspects related to the interaction of a *user* with software (Marcilly et al., 2015). The result of an interaction is described in terms of *effectiveness* and *efficiency* (Möller, 2017). A *usability evaluation* aims to evaluate the effectiveness and efficiency of user interactions with software (Bastien, 2010) managed by a *representative sample of test users*. A usability evaluation can improve the usability of a software during software development (Möller, 2017). Traditional usability evaluation methods, which encompass expert-based or user-based usability evaluation methods, can be applied to evaluate the usability of software (Jaspers, 2009). In the course of expert-based usability evaluation methods, experts put themselves in the role of prospective users evaluating an eHealth system. During heuristic evaluation, for example, heuristics are applied to evaluate the usability of a software (Nielsen, 1994). User-based usability evaluation methods integrate prospective users during the usability evaluation, typically achieved as *usability testing* in a usability laboratory (Borycki et al., 2013). During think aloud, test users that represent prospective users speak their thoughts out loud while performing tasks to evaluate the software's usability. The test users are observed by audio and video, usually by applying extensive technology in a usability laboratory (Kushniruk & Patel, 2004). Video observation requires a lot of resources, as extensive sequences of the test users are recorded (Stapelkamp, 2010). Traditional usability evaluations therefore consume a lot of time and resources (Pawson & Greenberg, 2009).

Failures in the use of eHealth systems identified via *eHealth evaluation* represent usability issues[1] of the eHealth system applied. Usability issues influence the interaction of users with the software (Marcilly et al., 2015). To avoid usability issues at an early stage of the *software life cycle*, an iterative approach to software development can be taken. Such an iterative approach is characterized by short iterations called sprints (Fuchs et al., 2013). During an iteration, a part of the eHealth system is optimized within a complete development cycle during the creation of the eHealth system, i.e., the software. A complete software development cycle comprises the definition of the requirements up to the examination of the finished software (Berg et al., 2014). An iterative approach is practiced within *agile development*. Two examples of agile development are extreme programming and

[1] In this book, the term usability issue is used interchangeably with the term usability problem.

scrum. Extreme programming was developed by Kent Beck and is characterized by an iterative-incremental approach to agile development (Obendorf et al., 2018). During extreme programming, small releases are made for the purpose of meeting changing customer requirements. Scrum is an agile project management method developed by Ken Schwaber and Jeff Sutherland (Rafi et al., 2015) that focuses on a software development project characterized by short sprints.

A basic premise of agile development is to quickly respond to changes in the requirements of a software such as an eHealth system. Further developments and modifications in the requirements of the software are implemented iteratively compared to traditional development, i.e., traditional software development. Traditional development is characterized by a plan-driven process of pre-defined phases (Fuchs et al., 2013). The result of one phase provides the basis for the subsequent phase. Prior to the start of a software development project, it is necessary to prepare the requirements and functional specifications of the software. Traditional software development accommodates summative usability evaluation, which refers to a final usability evaluation after the system has been developed (Joyce, 2019). In the course of iterative development, a formative eHealth usability evaluation is applied to identify usability problems in the early stages of software development (Broekhuis et al., 2021). Within agile development, the preparation of requirements and functional specifications of software is avoided because cooperation with the customer is valued more highly (Fuchs et al., 2013). This declaration is defined as one of four key aspects in the manifesto for agile software development (Berg et al., 2014).

The agile manifesto contains the fundamental ideas of agile development (i.e., agile software development). The idea of agile software development is expressed in four key values: "Individuals and interactions over processes and tools; working software over comprehensive documentation; customer collaboration over contract negotiation; responding to change over following a plan" (Beck et al., 2001). As an example, agile software development focuses on working software, not on the supporting documentation. To implement working software, it is necessary to consider usability as a part of quality in the early stages of the software life cycle (Möller, 2017).

However, traditional usability evaluation methods, such as user-based usability evaluation methods, are difficult to reconcile with the iterative development process of agile software development (da Silva et al., 2015). The iterative development process of eHealth systems requires the use of *eHealth usability evaluation methods* that can be deployed rapidly and easily. This requires new approaches in usability evaluation for eHealth systems, which can be realized through an *agile eHealth usability evaluation*.

1.2 Human–computer Interaction Relating to Universal Usability for eHealth

In this book, an agile eHealth usability evaluation focuses on generating early feedback from prospective users to enhance the usability of eHealth systems. The fields of usability and human-computer interaction are closely related (Hedvall, 2009) because both focus—in a broader context—on humans interacting with software. In the past, usability described a system's user-friendliness or ease-of-use (Spiliotopoulos et al., 2009). This definition no longer seems to be appropriate, as usability describes how a system can be utilized by users to create an effective, efficient, and satisfying user interface (Schoeberlein & Wang, 2011). To ensure the accessibility of the user interface of a system, it should be designed to be effective, efficient, and satisfactory for different users or groups of users (Moreno et al., 2009). Accessibility is an important consideration, as it is not only an issue for the elderly but also for disabled persons such as the visually impaired (Alajarmeh et al., 2011). Accessibility is the overarching effort to make products suitable for as many prospective users as possible (Nunes et al., 2012) who may be restricted, for example, by age-related declines or limited physical abilities.

Overall, accessibility can be considered a precondition for user perception and interaction with a system (e.g., web-based eHealth system), while usability improves the quality of user interaction with a system, an interaction that should be simple, efficient, and satisfying (Mori et al., 2011). Consideration of accessibility and usability in the development of an eHealth system can be achieved with user-centered design. Prospective users are involved in user-centered design through an iterative development process. By conducting eHealth usability evaluations during user-centered design, user feedback can be iteratively collected and considered in order to evolve eHealth systems. eHealth usability evaluations can be conducted with user-centered methods (used synonymously with usability evaluation methods) that are applied to understand human values, their needs, and their impact on interacting with a system (Oikonomou et al., 2009). One way to gather user feedback is by applying certain user-based eHealth usability evaluation methods (Jaspers, 2009; Moran, 2019). Empirical, qualitative data is collected through user-based eHealth usability evaluation methods such as usability testing. Quantitative data is typically collected through user-based eHealth usability evaluation methods such as questionnaires or online surveys. However, these traditional methods of eHealth usability evaluation are difficult to reconcile with an iterative development process. Therefore, the integration of agile, easily applicable, and useful eHealth usability evaluations into user-centered design is becoming increasingly important.

Agile user-centered design incorporates usability evaluation methods into the iterative development process, resulting in improved usability (Hussain et al., 2009a,b). In the course of agile software development, user feedback is obtained iteratively to improve the usability of the user interface. Integrating eHealth usability evaluation into agile development increases the usability of the evaluated system. In this way, the needs and special requirements of elderly users can be considered in the early stages of software development to enhance the usability of the eHealth system. Instead of referring to the software product as such, the experiences and needs of prospective users are included, which is why user-centeredness has become important in interaction design (Cao & Cheng, 2022). Interaction design also includes other concepts such as *user experience* and human-computer interaction. User-centered design and user experience extend beyond usability (Lewis, 2014). User experience focuses on the users' thoughts, feelings, and behavior when they engage with interactive systems (Soares et al., 2022) such as eHealth systems. User-centered design subsumes *usability engineering* (Lewis, 2014), which includes methods to achieve enhanced usability during the development of a new software (Richter & Flückinger, 2016). Human-computer interaction examines how computer technology (e.g., information and communication-based technology) affects human work and activity (Dix, 2009) and focuses on human values (Rogers, 2009).

The consideration of human values is essential if we want to achieve universal access to technology. Universal design originates in the conscious effort to meet the needs of an aging population and of disabled users to create an all-inclusive information society (Stephanidis & Akoumianakis, 2021). Society is changing with and through information and communication-based technologies (Utesheva & Boell, 2016); this is leading to new challenges in human-computer interaction because elderly users, who may suffer from age-related physiological, cognitive, or emotional declines (Garcia & de Lara, 2018), have special needs in terms of accessing and using a system (Leahy & Dolan, 2009). This book therefore addresses the agile usability evaluation of user-centered eHealth systems with focus on the special needs of elderly users in order to contribute to achieving universal usability in the information society.

1.3 Agility in Terms of eHealth and Software Life Cycle Models

Agile is understood as a methodology (Tsonev, 2021) or mindset (Schoor, 2022) to practice *agility*. Agility is a key issue in the context of eHealth, as the growing complexity of eHealth systems demands new approaches to usability evaluation that are agile, easy to apply, and useful as well as suitable for applications in health care. Agility is a basic premise of agile development, defined as the ability to quickly respond to changes in order to alter the requirements of a software such as an eHealth system.

A software is developed according to the stages[2] of the software life cycle, which can be divided into the following phases: requirements engineering, design, and evaluation (Magües et al., 2016). The sequence of individual phases to an overall process is described by the procedure model (Brandt-Pook & Kollmeier, 2008). Reasons for the applicability of the procedure model lie in the quality assurance and comparability of software development projects (Brandt-Pook & Kollmeier, 2008). Different procedure models can be applied to accomplish software development projects (Table 1.1).

Table 1.1 Models of the software development life cycle (Brandt-Pook & Kollmeier, 2008)

Waterfall Model	Verification Model	Spiral Model	Extreme Programming
Phase Model	Phase Model	Cyclic Model	Agile Model
The outcome of one phase flows into the following phase	Extension of the waterfall model to include software testing	The outcome of one cycle flows into the following iteration	The outcome of one iteration flows into the following iteration

The waterfall model consists of phases that follow one another. When one phase is completed, the next phase begins. For example, the software is not evaluated until it is fully implemented. The verification model is an extension of the waterfall model. The difference between these two models lies in their test culture. In the waterfall model, for instance, an acceptance test is performed after the requirements have been determined, up to and including the testing of individual program modules, which represent the smallest unit of a software (Brandt-Pook & Kollmeier, 2008). The verification model is suitable for large software development projects and is frequently used in regulated industry (e.g., the automotive or pharmaceutical industry). The spiral model is an extension of the previously mentioned models to include a cyclical approach. This means that the prototype is checked and tested in

[2] In the context of this book, the terms "stage" and "phase" of the software life cycle are used synonymously.

every cycle of the software development. In this context, a prototype is considered a software product. The final software is the sum of the individual software products.

An agile approach can also be applied to software development instead of the cyclic approach. Agile software development is based on the agile approach, which is based on the declarations written down in the agile manifesto. The user-oriented iterative approach to software development is one way to apply these principles.

Two well-known versions of agile software development are extreme programming and scrum. The extreme programming process model is characterized by short development cycles. Extreme programming is an agile software development framework that is characterized by an iterative, incremental approach to software development. Extreme programming involves small releases for the purpose of responding to changing customer requirements. A small release is the result of an iteration, which comprises a short, complete development cycle for a part of the software (Richter & Flückinger, 2016). The goal is to work on projects iteratively to respond more quickly to changes in software development (Hanser, 2010). In contrast, scrum is an agile project management method that was first developed by Ken Schwaber and Jeff Sutherland. In agile project management, the development of software is characterized by short iterations called sprints.

1.4 Research Gap and Research Questions

Combining the idea of agile software development with usability evaluation methods is possible in two ways (Sohaib & Khan, 2010). On the one hand, usability evaluation methods can be integrated into agile development in order to incorporate user feedback into the software development process at an early stage (Gundelsweiler et al., 2004). On the other hand, it is possible to integrate the idea of agile software development into usability evaluation methods (Pawson & Greenberg, 2009).

Agile user-centered design incorporates usability evaluation methods into the iterative development process, resulting in improved usability (Hussain et al., 2009a,b). There are numerous approaches to integrating usability evaluation methods into the concept of agile software development. For example, extreme programming (Gundelsweiler et al., 2004) or scrum (Singh, 2008) may be expanded with approaches on usability evaluation. This integration benefits from the improvement of software quality (Cavichi de Freitas et al., 2016). Integrating the idea of agile software development into traditional usability evaluation methods is difficult because traditional usability evaluation methods are cumbersome to apply, for instance due to the use of extensive technology (e.g., high-resolution equipment for extensive audio and video recordings applied in a usability laboratory). Certain methods from discount usability engineering, such as scenarios, simplified think aloud, and heuristic evaluation, adopt the idea of agile software development (Sinabell & Ammenwerth, 2022).

The concept of discount usability engineering, developed by Jacob Nielsen in the early 1990s, is a perfect fit for the health care setting, where the need for cost reduction is ever-present (Verhoeven & van Gemert-Pijnen, 2010) since significant costs are associated with the development of eHealth inventions (Sülz et al., 2021). However, only a few discount usability engineering approaches, such as rapid usability evaluation (Russ et al., 2010), low-cost rapid usability testing (Kushniruk & Borycki, 2017), or discount user-centered eHealth design (Verhoeven & van Gemert-Pijnen, 2010), are suitable for implementing an agile eHealth usability evaluation. Furthermore, existing approaches for agile usability evaluation are mostly based on expert-based usability evaluation methods (Hassan et al., 2019); user-based usability evaluation methods are mostly neglected due to their major investment in terms of time and effort. It is unclear, which usability evaluation approaches are appropriate and deployable to rapidly evaluate patient-centered eHealth inventions being iteratively developed in health care.

Furthermore, as mentioned previously the needs and special requirements of elderly users due to age-related declines are necessary to be considered in the early stages of software development to enhance the usability of the eHealth system. However, it is unclear whether it is possible (i.e., feasible) to conduct an easily applicable eHealth usability evaluation as well as counter and tackle challenges that might arise during the agile eHealth usability evaluation with elderly users.

This leads to the main research gap, which focuses on the suitability and deployability of existing approaches for agile eHealth usability evaluation applicable for the iterative development process of patient-centered eHealth inventions and the feasibility to conduct an easily applicable eHealth usability evaluation as well as counter and tackle challenges that might arise during an agile eHealth usability evaluation with prospective users, such as the elderly.

Based on this main research gap, four sub-research gaps can be outlined.

Sub-research gap 1:
A need exists to identify and prioritize applicable, rapidly deployable, and useful eHealth usability evaluation methods to support faster eHealth usability evaluations. This, in turn, results from a need to re-examine previously optimized eHealth usability evaluation methods (Wang et al., 2022) and from a lack of understanding as to which usability evaluation methods can be used specifically for the evaluation of eHealth inventions (Maramba et al., 2019). Most "summative usability tests do not consider eHealth-specific factors that could potentially affect the usability of

a system" (Broekhuis et al., 2021), making it necessary to clarify which eHealth specifics should be considered to conduct an agile eHealth usability evaluation.

eHealth usability evaluations are essential to the success of eHealth systems (Jaspers, 2009). eHealth usability evaluation "before, during, and after implementation of eHealth systems has been shown to be important and useful" (Parry et al., 2015) because the acceptance and use of health applications is hindered by poor design (Jaspers, 2009). Although the number of eHealth inventions is on the rise, the number of newly conducted eHealth usability evaluation studies is decreasing (Maramba et al., 2019). The usability of eHealth systems or their prototypes has been evaluated in the same way for decades (Broekhuis et al., 2019a). Existing usability evaluation methods have been evolved and optimized, making it necessary to re-examine the usability evaluation methods currently in use (Wang et al., 2022).

In eHealth software development projects, "both financial resources and sufficient development time" are scarce (Wildenbos, 2019a), making it necessary to integrate usability evaluation into the user-centered design process (Kushniruk & Borycki, 2017), which has the potential to improve the acceptance and security of eHealth systems such as electronic health records (Russ et al., 2010). Many usability evaluation methods, however, "require time and resources that an agile process cannot afford" (Magües et al., 2016).

Individual eHealth usability evaluation methods have been incorrectly named and are mis-referenced in the literature (Wronikowska et al., 2021), which is why it is necessary to identify equivalent terms used in literature for eHealth usability evaluation methods.

Sub-research gap 2:
The development of a toolbox suitable to implement agile, easily applicable, and useful eHealth usability evaluations is necessary, as there is a lack of existing approaches to eHealth usability evaluation toolboxes that support the selection of eHealth usability evaluation methods and which present both the strengths and weaknesses of the applicable methods in comparison. Recent research has indicated that there is a need to develop a toolbox to facilitate "the selection of an appropriate eHealth usability evaluation method" (Kip et al., 2022).

The selection of an appropriate usability evaluation method is often constrained by "practicability, accessibility of required human resources, and time to perform the evaluation study" (Peute et al., 2015). The development of proper toolboxes is recommended as an aid to the application of eHealth usability evaluation (Kip et al., 2022). The most popular eHealth usability evaluation methods are often the ones that are easy to apply, such as questionnaires (Broekhuis et al., 2019a). This shows the need to reconsider which eHealth usability evaluation methods are chosen, as studies indicate that questionnaires are not the best method for conducting an eHealth usability evaluation (Broekhuis et al., 2019a).

Sub-research gap 3:

As eHealth systems help the elderly to live independently, it is necessary to examine the applicability of eHealth usability evaluation methods and to review the feasibility of conducting an agile eHealth usability evaluation with elderly users (Sülz et al., 2021). However, health problems often discourage the elderly from using eHealth systems (Airola, 2021). Elderly people may suffer from age-related declines (Wildenbos et al., 2018) that affect the implementation of an agile eHealth usability evaluation and the relevant methods suitable for implementation. The aging population "could benefit from better communication with distant family and friends" as well as access to more information, including in health care (Leahy & Dolan, 2009).

Elderly users should be involved early in the software development process (Duque et al., 2019) to prevent eHealth systems from becoming too complex and unusable for this user group. For elderly persons who rarely use software, certain aspects may not seem as simple as they do for regular software users (Razak et al., 2013). Elderly users should be incorporated in the performance of eHealth usability evaluations because it is important to evaluate eHealth systems with representative test users (Wildenbos, 2019a). Suggestions have been made to use eHealth usability evaluation methods with elderly users, for instance by incorporating a relative (Wildenbos, 2019a); it is unclear, however, which method is applicable in a real-world environment, as adaption of an eHealth usability evaluation method to its prospective users is required in order to capture elderly people's needs (Isaković et al., 2016).

Current eHealth usability evaluations do not account for "variations in end-user populations and their effects on the perceived usability of a system" (Broekhuis et al., 2019b). "There is no specific guidance or information on how to implement an eHealth usability evaluation with elderly users", e.g., through remote usability testing (Hill et al., 2021).

Sub-research gap 4:

An investigation is needed of the challenges that may arise when conducting an agile, easily applicable, and useful eHealth usability evaluation with the elderly, and how to address and overcome these challenges, as there is a lack of guidance on conducting eHealth usability evaluations with this user group (Hill et al., 2021). Established approaches on usability evaluation do not, or only to a limited extent, consider the specific situation of elderly users. Design suggestions have been made for mHealth systems for elderly patients suffering from Alzheimer's disease or related dementias (Engelsma et al., 2021). However, there is a lack of useful recommendations for addressing challenges prior to, during, and after carrying out an agile eHealth usability evaluation with elderly participants. It is crucial to provide recommendations that consider different stages of the software life cycle, as eHealth usability evaluations have been shown to be important and useful "before, during, and after implementation of eHealth systems" (Parry et al., 2015).

Recommendations are relevant for conducting an agile usability evaluation with elderly users, as age-related declines make it more difficult to conduct an eHealth usability evaluation with this user group. For example, age-related declines "contribute to an accelerating deterioration in their health, with consequent loss of autonomy" (Cunha et al., 2014), which can cause challenges in conducting an eHealth usability evaluation if the elderly users are unable to visit a usability laboratory due to their limited mobility. Elderly persons often have problems completing simple tasks (Garcia & de Lara, 2018), such as reading a text, and "need an appropriate level of support, safety, and comfort" (da Silva et al., 2015). As elderly users find it difficult to embrace new technologies, it is crucial to consider the age-related declines of this population during the development of new technologies (Jakkaew & Hongthong, 2017).

Based on the sub-research gaps the aims of this book are derived as follows: (1) systematically identify and expert validate rapidly deployable eHealth usability evaluation methods to support faster eHealth usability evaluations, (2) develop a toolbox for agile eHealth usability evaluation, (3) examine the applicability of eHealth usability evaluation methods to elderly users and the feasibility of conducting an agile eHealth usability evaluation with elderly users, and (4) identify, dealing with, and overcome challenges that might be countered prior to, during, and after carrying out an eHealth usability evaluation with elderly users.

This book addresses improving the usability of patient-centered eHealth inventions for prospective user groups, such as the elderly, more broadly, and in so doing, contributes to universal usability by considering, addressing, and overcoming the challenges associated with conducting an agile, easily applicable, and useful eHealth usability evaluation with prospective users.

The research gap prompted the following research questions, which are divided into eight sub-research questions. Research questions 1 and 6 are each comprised of two sub-questions. Figures 1.1 to 1.4 show that the research questions are related to four research phases. Research phases 1 to 4 refer to the aforementioned sub-research gaps 1 to 4, respectively. Thus, for instance, research phase 1 is based on sub-research gap 1.

RESEARCH PHASE 1

Research aim

Systematically identification and expert-validation of rapidly deployable eHealth usability evaluation methods to support faster eHealth usability evaluations

Research questions

Main research question:
Which usability evaluation approaches are appropriate and deployable to rapidly evaluate patient-centered eHealth inventions being iteratively developed in health care, and is it feasible to conduct an easily applicable eHealth usability evaluation as well as counter and tackle challenges that might arise during the agile eHealth usability evaluation with elderly users?

First research questions
(RQ 1a):
Which eHealth usability evaluation methods are easily applicable, rapidly deployable, and useful to support faster eHealth usability evaluations?
(RQ 1b):
How can the identified eHealth usability evaluation methods be prioritized to achieve agile, easily applicable, and useful eHealth usability evaluations?
Second research question (RQ 2):
Which specific aspects for agile, easily applicable, and useful eHealth usability evaluations can be considered to evaluate eHealth inventions quickly?

Research methods

Step 1: Systematic literature review on eHealth usability evaluation methods
Step 2a: Expert-based iterative validation of eHealth usability evaluation methods to contrast expert knowledge with findings from literature to prioritize the previously identified eHealth usability evaluation methods (during step 1)
Step 2b: Developing of a checklist for conducting an agile, easily applicable, and useful eHealth usability evaluation

Overview of the most important result of research phase 1
▶ Forty-three eHealth usability evaluation methods were prioritized

Figure 1.1 Research aim as well as research questions related to the research phase 1. The research methods used are described for steps 1 to 2b, as well as an outlook on the most important research result of research phase 1

RESEARCH PHASE 2

Research aim

Development of a toolbox for agile eHealth usability evaluation

Research questions

Main research question:
Which usability evaluation approaches are appropriate and deployable to rapidly evaluate patient-centered eHealth inventions being iteratively developed in health care, and is it feasible to conduct an easily applicable eHealth usability evaluation as well as counter and tackle challenges that might arise during the agile eHealth usability evaluation with elderly users?

Third research question (RQ3):
What could a toolbox suitable to implement agile, easily applicable, and useful eHealth usability evaluations look like?

Research methods

Step 3: Based on the findings of step 1 and step 2 a toolbox consisting of rapidly deployable and useful eHealth usability evaluation methods was developed

Overview of the most important results of research phase 2

► Development of the ToUsE toolbox that comprises all 43 identified eHealth usability evaluation methods
► The toolbox offers method cards consisting of a method description, including references of each eHealth usability evaluation method
► The described eHealth usability evaluation methods are supplemented with information on aspects such as strengths, weaknesses, similarities, and/or relationships with other eHealth usability evaluation methods

Figure 1.2 Research aim as well as research questions related to the research phase 2. The research method used is described for step 3, as well as an outlook on the most important research results of research phase 2

RESEARCH PHASE 3

Research aim	Examination of the applicability of eHealth usability evaluation methods to elderly users and the feasibility of conducting an agile eHealth usability evaluation with elderly users
Research questions	***Main research question:*** Which usability evaluation approaches are appropriate and deployable to rapidly evaluate patient-centered eHealth inventions being iteratively developed in health care, and is it feasible to conduct an easily applicable eHealth usability evaluation as well as counter and tackle challenges that might arise during the agile eHealth usability evaluation with elderly users?
	Fourth research question (RQ4): Which eHealth usability evaluation methods are suitable for evaluating eHealth interventions with elderly users? ***Fifth research question (RQ5):*** Is it feasible to conduct an agile eHealth usability evaluation with elderly users in a real-world environment?
Research methods	***Conduction of a case study-designed, explorative study*** ***Step 4:*** Context-based selection of appropriate eHealth usability evaluation methods to implement the case study ***Step 5:*** Examination if the implementation of an agile, easily applicable, and useful eHealth usability evaluation is achievable with elderly users

Overview of the most important result of research phase 3

► Remote user testing combined with think aloud could be successfully applied to evaluate eHealth systems with elderly users
► However, not all established eHealth usability evaluation methods are suitable for conducting agile eHealth usability evaluations with elderly users

Figure 1.3 Research aim as well as research questions related to the research phase 3. The research methods used are described for step 4 and 5, as well as an outlook on the most important research result of research phase 3

RESEARCH PHASE 4

Research aim	Identification, dealing with, and overcome challenges that might be countered prior to, during, and after carrying out an eHealth usability evaluation with elderly users
Research questions	***Main research question:*** Which usability evaluation approaches are appropriate and deployable to rapidly evaluate patient-centered eHealth inventions being iteratively developed in health care, and is it feasible to conduct an easily applicable eHealth usability evaluation as well as counter and tackle challenges that might arise during the agile eHealth usability evaluation with elderly users?
	Sixth research questions (RQ 6a): What challenges may arise when conducting an agile, easily applicable, and useful eHealth usability evaluation with elderly users? ***(RQ 6b):*** How can the challenges that arise prior to, during, and after carrying out the eHealth usability evaluation conducted with elderly users be addressed and resolved?
Research methods	***Step 6:*** Systematic literature review on challenges of conducting agile, easily applicable, and useful eHealth usability evaluations with elderly users and how these challenges can be overcome

Overview of the most important result of research phase 4

▶ Twenty-four recommendations for addressing challenges prior to, during, and after carrying out the eHealth usability evaluation with elderly users were formulated

Figure 1.4 Research aim as well as research questions related to the research phase 4. The research methods used are described for step 6, as well as an outlook on the most important research result of research phase 4

Which usability evaluation approaches are appropriate and deployable to rapidly evaluate patient-centered eHealth inventions being iteratively developed in health care, and is it feasible to conduct an easily applicable eHealth usability evaluation as well as counter and tackle challenges that might arise during the agile eHealth usability evaluation with elderly users?

1a. Which eHealth usability evaluation methods are easily applicable, rapidly deployable, and useful to support faster eHealth usability evaluations?
1b. How can the identified eHealth usability evaluation methods be prioritized to achieve agile, easily applicable, and useful eHealth usability evaluations?
2. Which specific aspects for agile, easily applicable, and useful eHealth usability evaluations can be considered to evaluate eHealth inventions quickly?
3. What could a toolbox suitable to implement agile, easily applicable, and useful eHealth usability evaluations look like?
4. Which eHealth usability evaluation methods are suitable for evaluating eHealth interventions with elderly users?
5. Is it feasible to conduct an agile eHealth usability evaluation with elderly users in a real-world environment?
6a. What challenges may arise when conducting an agile, easily applicable, and useful eHealth usability evaluation with elderly users?
6b. How can the challenges that arise prior to, during, and after carrying out the eHealth usability evaluation conducted with elderly users be addressed and resolved?

1.5 Outline of this Book

Chapter 2 provides a description of the chosen research design, focusing on the triangulation study in relation to the research process and the research questions. The methodology used is presented in relation to research phases 1 to 4 of this book. To realize phase 1, a systematic literature review was conducted as step 1, expert interviews as step 2a, and two additional paper-based expert interviews as step 2b. Phase 2 reports on formalizing the identified and prioritized eHealth usability evaluation methods as step 3. Phase 3 explains the set-up, procedure, and data analysis of the explorative case study (step 4 and step 5). Phase 4 puts the focus on step 6, the implementation of the second systematic literature review.

Chapter 3 reports on the results of this book as they relate to the research questions in research phases 1 to 4. The chapter starts with an overview of the findings of paper I, paper II, and the research report. Phase 1 describes the findings of the systematic identification and expert validation of rapidly deployable eHealth usability evaluation methods to support faster eHealth usability evaluations. It reflects on agile, easily applicable, and useful eHealth usability evaluation methods to advance the ease-of-use of eHealth inventions. Phase 2 proceeds with an explanation of the developed toolbox for agile eHealth usability evaluations. The toolbox consists of rapidly applicable and potentially useful eHealth usability evaluation methods that can be used in a flexible and lightweight manner within a formative eHealth usability evaluation. Both phase 3 and phase 4 describe the findings of the explorative case study that was conducted with chronically ill Austrian elderly patients. Phase 3 explains the applicability of eHealth usability evaluation methods with elderly users and the possibility of implementing an agile eHealth usability evaluation with them. The case study showed that health-related barriers do not prevent elderly users from using eHealth systems. The implementation of an eHealth usability evaluation faces challenges which are finally described by phase 3 and phase 4. Phase 4 presents the recommendations to tackle those challenges that might be countered prior to, during, and after carrying out an eHealth usability evaluation with elderly users. The recommendations were synthesized from the findings of the exploratory case study and the second systematic literature review.

Paper I, paper II, and the research report are related to the methodology and to the results as shown in gray boxes at the beginning of research phases 1 to 4.

Chapter 4 starts by answering the research questions, beginning with the main research question, which describes the limitations, and moves on to new research questions. The research results are discussed considering the improvement of eHealth usability leading to increased patient-centeredness of eHealth inventions. The book ends with **Chapter 5**, which contains the conclusion, according to which the results of this book will facilitate the use of agile eHealth usability evaluations as a step towards universal usability of eHealth inventions.

The peer-reviewed publications and the research report can be found in the ESM of this book.

Method 2

A triangulation study was conducted combining iterative expert interviews with an exploratory case study. The main method of this study is qualitative and was triangulated with other qualitative methods to ensure the accuracy of the results. During research phase 1, a systematic review combined with expert interviews was carried out using inductive and deductive qualitative content analysis. In the course of research phase 2, method cards were developed to formalize the usability evaluation methods for evaluating eHealth usability. During research phases 3 and 4, an explorative case study supplemented with a systematic literature review was performed.

2.1 Research Design

The research design is defined as "the logic that links the data to be collected (and the conclusions to be drawn) to the initial questions of study" (Yin, 2003). The research design can thus be understood as the logic by which the collected data is related to the research question. This study primarily collected qualitative data from human-computer interaction specialists and the participants gained within the case study. To answer the research question, the choice of the research method – qualitative or quantitative – is required (Kuckartz, 2014). In addition to the primarily qualitative data, quantitative data was collected as well (within the prioritizations and participants' ratings).

Supplementary Information The online version contains supplementary material available at https://doi.org/10.1007/978-3-658-44434-1_2.

I. Sinabell, *Agile eHealth Usability Evaluation*,
https://doi.org/10.1007/978-3-658-44434-1_2

The triangulation study was chosen by combining qualitative research methods because it allows for “confirmation of findings, more comprehensive data, increased validity”, and a better understanding of the phenomena being studied (Bekhet & Zauszniewski, 2012). In the context of this book, the phenomenon under examination refers more broadly to human-computer interaction, which provides the fundamentals for implementing a usability evaluation. The main research question of this book is to explore which usability evaluation approaches are appropriate and deployable to rapidly evaluate patient-centered eHealth inventions being iteratively developed in health care, and if it is feasible to conduct an easily applicable eHealth usability evaluation as well as counter and tackle challenges that might arise during the agile eHealth usability evaluation with elderly users. To address the suitability of an agile, easily applicable, and useful usability evaluation within the iterative development process of patient-centered eHealth inventions that is applicable to the elderly, it is necessary to examine, by extension, the behavior of elderly users during eHealth usability evaluations to gain insight into likely challenges in applying eHealth usability evaluation methods to such users.

Qualitative research was chosen for the purpose of answering the research questions because it pertains to “the set of meanings, values, beliefs, and social behaviors that would not be quantifiable” (da Silva Santos et al., 2020). The use of qualitative research methods in this book is justified to gather experts’ experiences and opinions on established usability evaluation methods and elderly users’ behavior during eHealth usability evaluation. The rationale for choosing a qualitative research method was to gather information about experiences, specialized knowledge, and the perceptions of experts by analyzing verbal expressions (Helfferich, 2011). Experts have specialized knowledge that serves to answer the research question (Bogner et al., 2002). The iteratively conducted expert interviews represent a systematized semi-structured expert interview, as the focus is on a comprehensive survey of the experts’ specialized knowledge (Bogner et al., 2014).

In this book, the triangulation study according to the authors’ Schreier & Odağ (2010) definition represents a multiple method study because several research methods were chosen, which are subject to the same paradigm, the qualitative research. Triangulation implies that more than one perspective is used to investigate the research question, thus increasing confidence in the validity of the results (Kuckartz, 2014). Perspective refers here to the chosen method, which can be qualitative or quantitative. A quantitative method was applied within content analysis to analyze the experts’ prioritization. The experts either recommended or did not recommend the use of an eHealth usability evaluation method. Those eHealth usability evaluation methods that were neither recommended nor not recommended by the experts were considered potentially useful for the evaluation of an eHealth invention.

A combination of both qualitative research methods and the quantitative procedure of the frequency rating within the inductive and deductive content analysis was thus chosen to strengthen the accuracy of the results. As a further quantitative aspect of this research design, elderly users' ratings of identified usability issues of the web-based eHealth invention were counted during the explorative case study. The number of usability issues encountered in each session of the eHealth usability evaluation was counted for each elderly user. The triangulation study was also selected to obtain new research results in a complementary manner (Ammenwerth et al., 2003), such as contrasting the experts' prioritization to the results of the systematic literature review or supplementing the case study results with the systematic literature review to derive recommendations on how to address challenges prior to, during, and after carrying out the eHealth usability evaluation conducted with elderly participants.

Both systematic literature reviews were chosen to "identify gaps, deficiencies, and trends in the current evidence" (Munn et al., 2018a). The systematic literature reviews were applied within this book to "confirm current practice", "identify and investigate conflicting results", and to "identify and inform areas for future research" (Munn et al., 2018b). A search strategy which included such inclusion and exclusion criteria was selected to comply with the structured procedure of the systematic literature search. The study selection procedure, including and excluding studies, was visualized with the PRISMA flow diagram (Page et al., 2021).

Figures 2.1 and 2.2 show the chosen research design in relation to the research questions. Overall, the triangulation study consisted of two systematic literature reviews, phone-based as well as paper-based qualitative expert interviews, formalization of the eHealth usability evaluation methods, and an explorative qualitative case study. The transcribed text material was analyzed using quantitative inductive and deductive content analysis by counting the frequencies of expert-rated prioritization (recommended, not recommended, and potentially useful) of eHealth usability evaluation methods. As a first main step, a systematic review was conducted complementary to the expert-based validation of eHealth usability evaluation methods. As a second main step, a second systematic literature review was conducted supplementary to the case study to strengthen the results triangulatively.

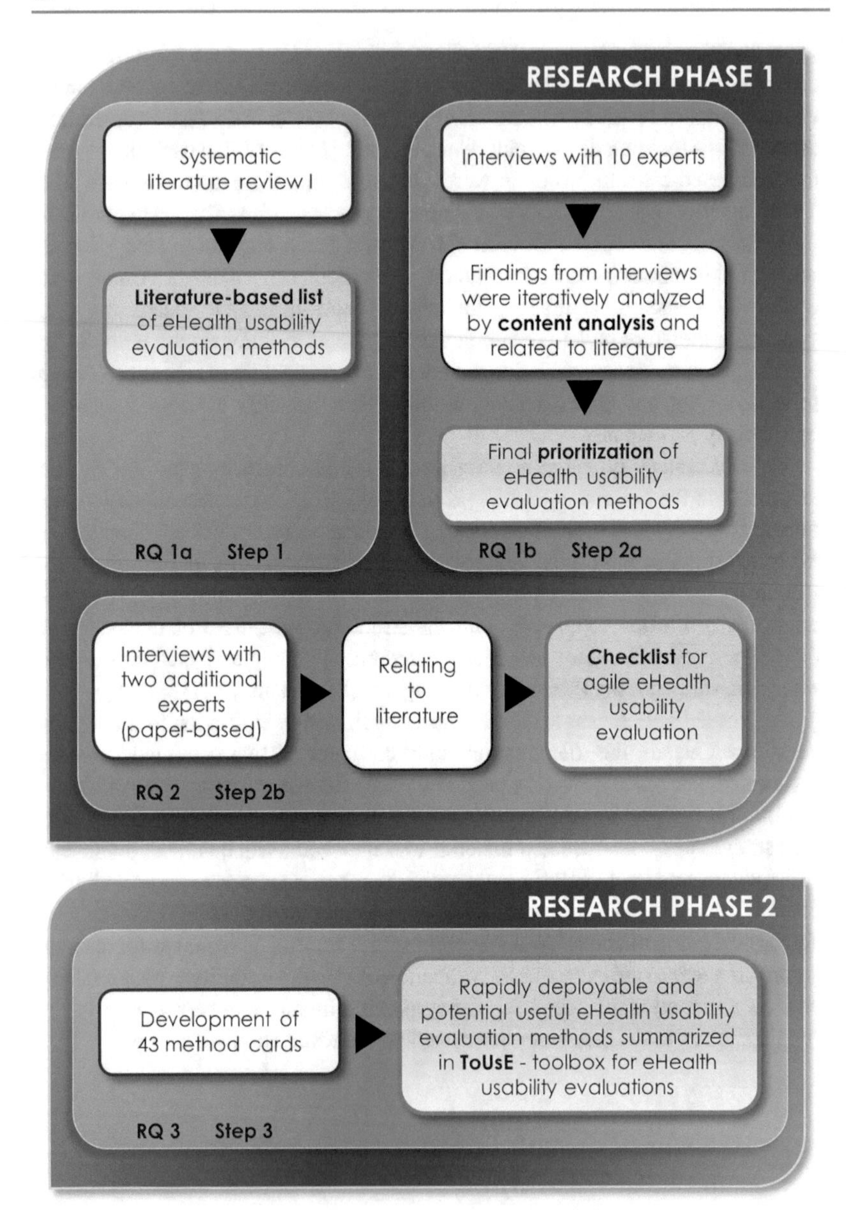

Figure 2.1 Research process relating to research questions 1 to 3. Boxes with gray shading indicate results of each research question. The black arrows represent the direction of the research process (RQ = research question)

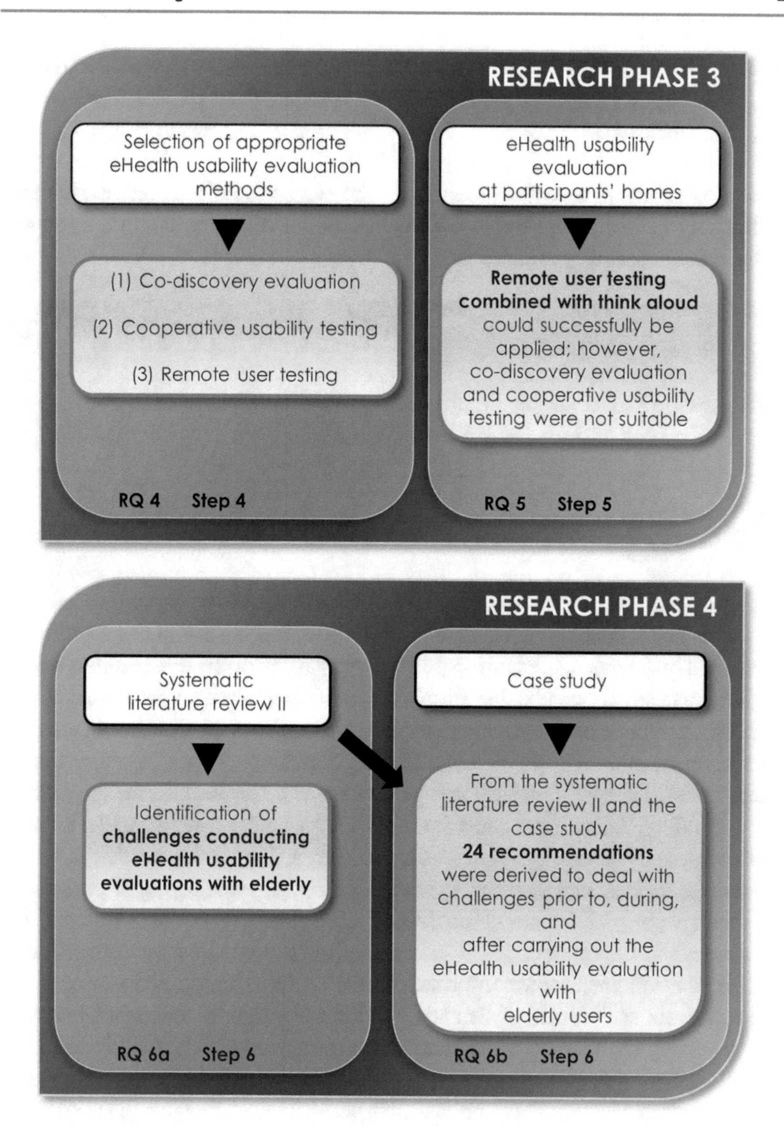

Figure 2.2 Research process relating to research questions 4 to 6. Boxes with gray shading indicate results of each research question. In the course of phase 4, formulated recommendations are synthesized from the case study and the systematic literature review II is marked with a large black arrow. The small black arrows represent the direction of the research process (RQ = research question)

2.2 Research Phase 1: Identifying and Expert-validating Rapidly Deployable eHealth Usability Evaluation Methods

The research aim of research phase 1 was the systematic identification and expert validation of rapidly deployable eHealth usability evaluation methods to support faster eHealth usability evaluations.

> The method of research phase 1 has been partially published in *Applied Clinical Informatics*, an international journal published by Thieme.
> **Sinabell, I.**, & Ammenwerth, E. (2022). Agile, Easily Applicable, and Useful eHealth Usability Evaluations: Systematic Review and Expert-Validation. *Applied Clinical Informatics*, 13(01), 67–79.
> https://doi.org/10.1055/s-0041-1740919

Step 1—Systematic literature review on eHealth usability evaluation methods
A systematic literature review within the fields of eHealth, usability, and agility was conducted to develop a literature-based list of usability evaluation methods that are potentially useful for conducting an eHealth usability evaluation. Search terms were selected from the areas of eHealth, usability, and agility such as eHealth, telemonitoring, telehealth, mHealth, "mobile health", "electronic health", health, "clinical informatics", medical, "medical computer science", "health information technology", usability, "human-computer interaction", "usability testing", agile, extreme, rapid, fast, and iterative combined with evaluation, framework, model, approach, process, processes, concept, testing, development, or engineering. To accommodate the specifics of the search engines of each database, the search terms were adjusted where appropriate. For instance, Medical Subject Heading (MeSH) terms such as telemedicine, medical informatics, user-centered design, and user-computer interface were used for the systematic literature search in Medline (via PubMed). The following databases were selected for the search: ACM Digital Library, IEEE Xplore, and Medline (via PubMed). To consider emerging eHealth usability evaluation methods that have not been published in peer-reviewed literature, the systematic literature search was supplemented with gray literature from GoogleScholar, in the course of which the first 30 pages of results were considered.

A total of 3,981 results (peer-reviewed and non-peer-reviewed literature) met the inclusion criteria. To include a paper in the review process, the paper had to contain a description of an eHealth usability evaluation method, report on agility as a way to rapidly implement eHealth usability evaluations, and relate to the applicability of eHealth usability evaluation methods. The search was limited to English-language papers published from January 2008 to June 2019. The systematic literature review focused on obtaining descriptions of potentially useful eHealth usability evaluation methods. Papers consisting of reports, conference posters, or presentations were excluded. After applying the search queries, all findings were downloaded into Zotero reference manager to review them for possible inclusion (n=3,657). The inclusion criteria were: (1) relevance to the three thematic areas (eHealth, usability, and agility) of this systematic literature review, (2) description of an eHealth usability evaluation method, (3) eHealth usability evaluation methods that emphasize the evaluation stage of the software life cycle, and (4) papers that were peer-reviewed and non-peer-reviewed. The exclusion criteria were: (1) papers not published in English, (2) papers that do not emphasize eHealth systems being evaluated, (3) no description of eHealth usability evaluation methods potentially useful to implement an agile, easily applicable, and useful eHealth usability evaluation, and (4) no information that the described eHealth usability evaluation method can be rapidly deployed. In this context, eHealth usability evaluation methods encompass a method, model, approach, process, or concept that can be rapidly deployed to implement an agile, easily applicable, and useful eHealth usability evaluation.

The findings were analyzed according to a two-step procedure as follows: (1) screening the papers based on title and abstract that matched the inclusion and exclusion criteria and (2) reading the full text of those papers that fully met the inclusion criteria. If an eHealth usability evaluation method was theoretically or practically integrated in an agile, easily applicable, and useful eHealth usability evaluation, the eHealth usability evaluation method was considered potentially useful for eHealth usability evaluation. From these included papers (n=287), a list of 29 eHealth usability evaluation methods was extracted that were potentially useful to conduct an agile, easily applicable, and useful eHealth usability evaluation. This list of potentially useful eHealth usability evaluation methods was supplemented by an ongoing hand search in peer-reviewed journals simultaneously with the interviews with experts (step 2) to include recently published literature. In addition, the snowballing approach was applied, and findings for additional relevant literature were examined.

Step 2a—Expert-based iterative validation of eHealth usability evaluation methods

The second step involved the validation of the extracted list of 29 potentially useful eHealth usability evaluation methods by experts. The validation was done iteratively, through continuous validation of each eHealth usability evaluation method identified in step 1 by ten experts contrasted with literature. Five iterations were carried out, each with two different experts. During each iteration, (1) semi-structured interviews with two experts were carried out subsequently to each other and these findings were (2) compared with the literature in order to newly add, alter, or remove potentially useful eHealth usability evaluation methods. The experts' statements were compared with the literature to ensure that the most recent eHealth usability evaluation methods were considered. The interviews with the experts were conducted iteratively, which was motivated by the iterative approach of software development.

The usability experts were identified through professional associations. In selecting the experts, evidence of at least ten years of professional experience in the field of usability, user experience, and/or agile software development as well as a specific job title were necessary (Table 2.1). The semi-structured interviews were based on an interview guide that was prepared in advance (Table A, available in the ESM). The interview guide was designed for a half-hour interview that consisted of two thematic areas: (1) expert opinion on which eHealth usability evaluation methods are rapidly deployable and can be used for agile, easily applicable, and useful eHealth usability evaluation and (2) expert opinion on which of the potentially useful eHealth usability evaluation methods identified in step 1 are recommended, potentially useful, or not recommended for use in agile, easily applicable, and useful eHealth usability evaluations.

Table 2.1 Inclusion criteria of professional activity for expert interviews

usability engineer, usability or user experience specialist, experience consultant, user experience architect/designer, usability, interaction, or product designer

The ten interviews with the usability experts were conducted in March and April 2020. The experts were contacted via email. The majority of experts responded within 24 hours of being contacted (7/10). Two expert interviews were scheduled that same day, and five more experts agreed to participate a day later. Three more experts agreed to the interviews six, ten and eleven days, respectively, after being contacted by email (Table 2.2).

Table 2.2 Duration of response after contacting the experts (n=10)

period	number of experts	period	number of experts
Date set for the expert interview		**Date of expert interviews**	
the same day	2	the same day	1
one day later	5	one day later	–
6 days later	1	2 to 6 days later	4
7 days later	1	7 days later	1
10 days later	1	10 days later	1
11 days later	–	11 days later	2
12 days later	–	12 days later	1

Note: The dates for the two additional paper-based expert interviews (step 2b) were agreed on day 11 and 12, respectively

All ten interviews were transcribed with a total of 29,799 words, which corresponds to 42 pages (Table 2.3). The interviews were analyzed using a combination of inductive and deductive content analysis. The inductive categorization of the qualitative content analysis following Mayring (2015) was chosen to form selective categories (one category corresponds to one eHealth usability evaluation method) and to prescribe the level of abstraction. The interview transcripts were used for inductive content analysis, while the literature-based list of potentially useful eHealth usability evaluation methods was used for deductive content analysis, with each eHealth usability evaluation method assigned to a pre-defined category.

Table 2.3 The number of transcribed words and pages

Interview with	Transcribed words	Transcribed pages
Expert 1	1,148	3
Expert 2	984	3
Expert 3	3,282	6
Expert 4	3,381	7
Expert 5	1,586	4
Expert 6	3,209	6
Expert 7	11,775	4
Expert 8	918	2
Expert 9	2,111	4
Expert 10	1,405	3
Total	*29,799*	*42*

The analysis consisted of two steps as follows: (1) each eHealth usability evaluation method from the literature-based list (identified in step 1) was counted as recommended or not recommended by the experts. The number of recommendations (as well as non-recommendations) made by the experts were documented for each eHealth usability evaluation method. The absolute frequencies of the number of recommendations made by the experts either in favor of or against an eHealth usability evaluation method were used to analyze the data; (2) individual eHealth usability evaluation methods were summarized using the same methodology. This was done for both eHealth usability evaluation methods that were recommended as well as for methods that were not recommended.

For instance, both asynchronous usability testing and unmoderated usability testing are conducted automatically without an evaluator. Since these eHealth usability evaluation methods can be considered the same, they can be subsumed under the term unmoderated usability testing.

The expert interviews were completed after five iterations because the findings were saturated after the fourth iteration. Saturation was achieved when the number of newly added, altered, or removed eHealth usability evaluation methods no longer changed. After the fourth iteration, there was a sharp decline in the number of newly added, altered, or removed eHealth usability evaluation methods (Table 2.4).

Table 2.4 Saturation of information content of interviews with experts

Iteration	Value
Iteration 1	17
Iteration 2	11
Iteration 3	4
Iteration 4	1
Iteration 5	1

Note: Value refers to the absolute number of newly added, recommended, and not recommended eHealth usability evaluation methods

The absolute numbers of the eHealth usability evaluation methods that were recommended or not recommended by the experts are presented in the Tables B and C (available in the ESM).

To protect the privacy of the experts, no transcription protocols are published within the scope of this book. The number of decisions made by the experts in favor of or against an eHealth usability evaluation method was counted from the anonymously transcribed protocols, making it impossible to draw conclusions about which individual expert made which statements.

Step 2b—Developing a checklist for conducting an agile, easily applicable, and useful eHealth usability evaluation

As a secondary outcome of step 2, expert recommendations for conducting an agile eHealth usability evaluation were derived. To strengthen these expert recommendations, two additional expert interviews were conducted. In the course of the two additional expert interviews, expert opinion was sought on key aspects (e.g., number of test participants, documentation, hardware or software used, time required to conduct an agile eHealth usability evaluation) that may influence the implementation of an agile eHealth usability evaluation.

Two paper-based interviews with experts were conducted subsequently to the expert interviews in step 2a. The interview questions used to conduct the expert interviews represent an evolvement of the semi-structured interview guide applied in step 2a. Two additional questions focused on aspects that should be considered when conducting an agile eHealth usability evaluation, such as the number of test participants, documentation, hardware or software used, or the time required to complete the eHealth usability evaluation. The experts' opinion served as a basis for developing a checklist for agile eHealth usability evaluations. To obtain the opinion of the experts on the significance of a key aspect, the experts were asked (1) which aspects are necessary to consider prior to the implementation of an eHealth usability evaluation to improve the preparation of agile eHealth usability evaluations, and (2) which aspects could be adjusted so that an eHealth usability evaluation can be conducted quickly? (Table D, available in the ESM).

Two experts were interviewed independently from each other. To exclude mutual influence of the experts, both interviews were conducted separately. The questions were emailed to the experts, with a request that they answer the questions in writing to avoid influence from the researcher (author of the book) or the other respondent.

The paper-based interviews were analyzed via inductive content analysis. Since two expert assessments were made, no frequencies of aspects were counted. The identified aspects were linked to the results of step 2 in order to find indicators on aspects that should be considered when conducting an agile eHealth usability evaluation. As an example, one expert recommended reducing various distractions (e.g., ringing cell phones) during eHealth usability evaluation to enable rapid implementation. Therefore, this recommendation was included as an item in the checklist.

A sample of a written informed consent signed by all experts (step 2a and step 2b) is documented in the ESM (Table E). To protect the experts' privacy, the signed written informed consents are not published as part of this book.

2.3 Research Phase 2: Development of a Toolbox for eHealth Usability Evaluations

The research aim of research phase 2 was the development of a toolbox for agile eHealth usability evaluation.

The method of research phase 2 is partially described in the following research report: **Sinabell, I.**, & Ammenwerth, E. (2022). ToUsE: Toolbox for eHealth Usability Evaluations. Austria, Hall in Tirol: UMIT TIROL.

Step 3—Developing a toolbox consisting of rapidly deployable and useful eHealth usability evaluation methods
Starting from step 1 and step 2, all identified eHealth usability evaluation methods were formalized by creating method cards. The method cards include the name of the eHealth usability evaluation method, the number of the method card, a method description, notes on related method cards, if applicable, and references. The method cards consist of a front and back side and are designed to be cut out and used prior to the agile eHealth usability evaluations. The relevant literature (identified in research phase 1, step 1) was used to create a comprehensive overview of expert-based and user-based eHealth usability evaluation methods.

The described eHealth usability evaluation methods are supplemented with information on strengths, weaknesses, similarities, and/or coherence to other eHealth usability evaluation methods described in the toolbox.

To provide a comprehensive description, the snowball principle was applied, and a hand search was conducted to identify even more relevant literature. Based on the results of the systematic literature search in step 1, 66 additional papers reporting on usability evaluation methods that are rapidly deployable and applicable in the field of eHealth were identified. If the paper reported on the strengths and weaknesses of the described eHealth usability evaluation method, this information was extracted and provided as additional information below the method cards. Particular attention was paid to recently published literature.

The toolbox aims to relate single eHealth usability evaluation methods to each other. For example, the authors Marcilly et al. (2021) report on an eHealth usability evaluation method called "competitive usability evaluation" that is useful for comparing the usability of electronic health records. This eHealth usability evaluation method was not identified as part of the expert-based validation because the publication does not give any indication whether the method is rapidly deployable. For this reason, competitive usability evaluation is described as an alternative eHealth usability evaluation method to A/B testing.

2.4 Research phase 3: Applicability of Agile eHealth Usability Evaluation with Elderly Users

The research aim of research phase 3 was the examination of the applicability of eHealth usability evaluation methods to elderly users and the feasibility of conducting an agile eHealth usability evaluation with elderly users.

The method of research phase 3 has been published in *Universal Access in the Information Society*, an international journal published by Springer.
Sinabell, I., & Ammenwerth, E. (2022). Challenges and Recommendations for eHealth Usability Evaluation with Elderly: Systematic Review and Case Study. *Universal Access in the Information Society*. https://doi.org/10.1007/s10209-022-00949-w

An explorative case study was conducted to examine if an agile eHealth usability evaluation can be conducted with the elderly in a real-world environment. In the course of the case study, an eHealth usability evaluation of an eHealth invention was conducted. The eHealth invention is a web-based eHealth system for patient-physician communication that allows patients to retrieve diagnostics reports in a decentralized manner. In the course of this book, an eHealth system that targets patients was chosen to conduct a case study, as these technologies have the potential to support patient-physician interaction (Nunes et al., 2019). The prospective user group of the web-based eHealth invention are elderly patients who frequently suffer from diseases or are chronically ill.

A pretest with an elderly person was conducted prior to the case study (March 2021). The pretest was used to rehearse the procedure and technical equipment used in the agile eHealth usability evaluation. After the pretest was completed, the eHealth usability evaluation was conducted with six additional elderly people. The pretest and the case study were conducted in March 2021. The median age of the elderly users was 70 years, and each session of the eHealth usability evaluation lasted a median of 27 minutes.

Step 4—Context-based selection of appropriate eHealth usability evaluation methods to implement the case study

To conduct the eHealth usability evaluation, (1) appropriate eHealth usability evaluation methods for eHealth usability evaluation were selected and (2) inclusion and exclusion criteria were defined to recruit elderly participants.

Three eHealth usability evaluation methods for evaluating eHealth usability were selected: (1) co-discovery learning, (2) cooperative usability testing, and (3) remote user testing combined with think aloud. These three eHealth usability evaluation methods were selected from the previously developed toolbox for eHealth usability evaluations (ToUsE).

Co-discovery evaluation means that two (or more) participants work on tasks simultaneously (Bastien, 2010). Research has shown that elderly users prefer to complete tasks together or with caregivers (Wildenbos, 2019a). Co-discovery evaluation was chosen because elderly users should work together while performing tasks during the eHealth usability evaluation.

Cooperative usability testing means that a video recording is made during the eHealth usability evaluation, which is viewed together with the elderly user after the eHealth usability evaluation has been conducted (Frøkjær & Hornbæk, 2005). Cooperative usability testing was selected to combine domain expertise with human-computer interaction expertise because of the collaboration between the elderly and the researcher (author of the book).

Remote user testing means that the evaluator and participants are in different locations (Bastien, 2010). During think aloud, participants express their thoughts aloud while performing tasks (Jaspers, 2009). Remote user testing combined with think aloud was chosen to give elderly users the opportunity to conduct the eHealth usability evaluation in their homes.

The intention was to use all three selected eHealth usability evaluation methods for eHealth usability evaluation with all elderly participants (n=7). Therefore, all seven elderly patients were invited to participate in eHealth usability evaluation conducted with each selected eHealth usability evaluation method.

Elderly participants

The study team consisted of a researcher (author of the book), two software developers who developed the web-based eHealth invention, and elderly patients from Austria as participants.

Formative usability evaluation "moved toward the model of five to seven representative users generally finding about 80% of the" usability issues (Downey, 2007). Therefore, seven elderly participants were chosen to conduct the eHealth usability evaluation. Saturation was achieved after conducting the eHealth usability evaluation with the fourth elderly participant, as the usability issues found by the elderly repeated. The residence of the elderly participants was restricted to Austria. The eHealth usability evaluation was conducted in the elderly persons' homes due to their limited mobility.

The inclusion criteria for the elderly participants were as follows: (1) age of 60 years or above, (2) living independently, (3) chronically ill or suffering from a disease, (4) email address, (5) basic knowledge of how to use a computer, (6) ownership of their own computer or technical equipment, and (7) ability to understand and sign the written consent form.

The exclusion criteria were as follows: (1) no email account, (2) residence outside of Austria, (3) no ability to understand and adequately express themselves in the German language, and (4) lack of signed written informed consent.

German language proficiency was required to complete the information consent to conduct the eHealth usability evaluation.

Step 5—Examination of the implementation of an agile, easily applicable, and useful eHealth usability evaluation is achievable with elderly users

First, the procedure of an eHealth usability evaluation was explained to the elderly participants. In addition, they were asked to sign a written informed consent form. The elderly participants were informed that the web-based eHealth invention and not their own performance was the subject of evaluation. The eHealth usability evaluation was conducted in March 2021. The entire design of the eHealth usability evaluation was tailored to the participants' physical and cognitive abilities. Prior to the start of the eHealth usability evaluation, the elderly participants were asked whether they would prefer to use their own computer or a computer provided by the researcher (author of the book). Short tasks were developed to help the elderly participants concentrate during the eHealth usability evaluation. To accomplish the development of the tasks, a fictional scenario was defined. Sufficient time was scheduled to explain the tasks to the elderly participants prior to the eHealth usability evaluation.

Scenario and tasks for eHealth usability evaluation

The three selected eHealth usability evaluation methods—remote user testing combined with think aloud, cooperative usability testing, and co-discovery learning—were applied in a different order with each participant to reduce the risk that the results of the eHealth usability evaluation might influence each other because of their consecutive execution. A scenario was formulated, and three tasks were derived from the scenario. The scenario assumed that the elderly participant would receive a prescription via email and in return had to provide personal information (email address and phone number) to the physician. After receiving the email notification, the prescription would then be accessible via the web-based eHealth invention.

Prototype of the web-based eHealth invention

Initially, a Transaction Authentication Number (TAN) was sent to the telephone number of the elderly participant. In the first step, the elderly participant had to enter the received TAN into the input field in the front end of the web-based eHealth invention. After verifying the TAN, the prescription was available for downloading in a second step. The third step after receiving the prescription offered two options: (1) managing personal data by deleting the personal data directly (otherwise the personal data would be deleted automatically after ten days) or (2) closing the front end of the web-based eHealth invention. In a fourth step, a final optional evaluation of the prescription transmission process was available.

The following scenario was developed: "A patient suffers from back pain but is hesitant to visit the doctor's office because of the serious COVID situation. The patient contacts the doctor by phone. Four physical therapy sessions are prescribed by the physician in consultation with the patient. The patient receives the prescription through the web-based eHealth intervention and does not have to visit the physician in person. This allows the patient to visit the physical therapist directly". Based on the scenario, the following tasks were derived: (1) the patient must find the contact information for the physician in the email received, (2) the patient must save the prescription locally on the computer and send it to the printer, and (3) after saving the prescription on the computer, the participant must delete the email address and other personal information.

Data analysis of eHealth usability evaluation

Qualitative notes from the usability sessions were systematically analyzed. The number of usability issues described and stated by the elderly participants was systematically counted. No inference can be made about an individual elderly participant since the written notes have been anonymized. Some usability issues were mentioned multiple times by the elderly participants. For example, some of the elderly users indicated that they had difficulty distinguishing the color of buttons. The usability issue "difficulty distinguishing color of buttons" was mentioned during two sessions of the eHealth usability evaluation, each of which was conducted with one elderly user. Therefore, this usability issue was counted twice.

To protect the participants' privacy, the original notes that were analyzed were not published as part of this book. A sample of a written informed consent form the elderly participants is documented in the ESM (Table F).

2.5 Research phase 4: Challenges of Conducting an Agile eHealth Usability Evaluation with Elderly Users

The research aim of research phase 4 was to identify, deal with, and overcome challenges that might be countered prior to, during, and after carrying out an eHealth usability evaluation with elderly users.

The method of research phase 4 has been published in *Universal Access in the Information Society*, an international journal published by Springer.
Sinabell, I., & Ammenwerth, E. (2022). Challenges and Recommendations for eHealth Usability Evaluation with Elderly: Systematic Review and Case Study. *Universal Access in the Information Society.*
https://doi.org/10.1007/s10209-022-00949-w

Step 6—Systematic literature review on the challenges of conducting agile, easily applicable, and useful eHealth usability evaluations
The explorative case study was conducted to learn about the challenges of applying eHealth usability evaluation methods and how these challenges might be overcome. Subsequently to the case study, a systematic literature review was conducted to classify challenges according to age-related barriers such as cognition, motivation, perception, and physical abilities.

The systematic literature was conducted to: (1) identify papers that report on the use of eHealth evaluation methods for elderly and (2) identify the challenges of conducting agile, easily applicable, and useful eHealth usability evaluations with elderly users. The systematic literature review was conducted in March 2022.

Four databases were searched for relevant articles: ACM Digital Library, Cochrane Library, IEEE Xplore, and Medline (via PubMed). The following search terms were used that related to the four topic areas of this study (usability, eHealth, elderly, and challenges): eHealth, mHealth, "medical informatics applications", elderly, elder*, senior, old, usability, "usability testing", challenges, limitations, feasibility, acceptability, and usability. While searching Medline (via PubMed), MeSH terms such as telemedicine, aging, and user-centered design were also considered.

To ensure that no relevant literature was missed, different combinations of search terms were used depending on the database search.

Papers were included that (1) were published in the last ten years (2012–2022), (2) focused on eHealth interventions, and (3) reported challenges in conducting eHealth usability evaluations with elderly users. To obtain a large number of relevant papers, the search was extended to include papers that report not only on challenges, but also on the acceptability, applicability, or feasibility of eHealth usability evaluation with elderly users. Findings on challenges with disabled participants were also included to avoid excluding relevant information from this related field. Papers were excluded according to the following criteria: (1) non-peer-reviewed papers (other than conference papers), (2) papers that did not focus on eHealth interventions, and (3) papers that were not written in English.

The following information was extracted from these papers: (1) applied eHealth usability evaluation methods for evaluating eHealth usability with elderly users (for all 20 papers) and (2) challenges in conducting eHealth usability evaluations with elderly users, if any.

The search was not limited to papers that focus exclusively on user-based eHealth usability evaluations because the search subject also included challenges to eHealth usability evaluations in which usability was evaluated by experts (expert-based eHealth usability evaluations). All retrieved papers were imported into the JabRef reference manager and duplicates were removed (n=21). The titles and abstracts of the papers were reviewed based on relevance to the defined inclusion and exclusion criteria (n=279). Based on the titles and abstracts, 243 papers were excluded; in the full text review, the remaining papers were included (n=36). Of these 36 papers, 20 were identified that reported on the use of eHealth usability evaluation methods with elderly users. Three additional papers were identified through hand search that reported on the challenges of evaluating the usability of eHealth among elderly users.

Results 3

The contents of the two peer-reviewed articles, paper I and paper II, which form the main part of the book, are outlined below.

Paper I:
Sinabell, I., & Ammenwerth, E. (2022). Agile, Easily Applicable, and Useful eHealth Usability Evaluations: Systematic Review and Expert-Validation. *Applied Clinical Informatics*, 13(01), 67–79.
https://doi.org/10.1055/s-0041-1740919

During the iterative approach, expert knowledge was contrasted with findings from the systematic literature review. Forty-three eHealth usability evaluation methods were systematically identified and assessed regarding their ease-of-use and usefulness for conducting eHealth usability evaluations through expert interviews with ten usability experts. After literature review and expert interviews, ten recommended eHealth usability evaluation methods, 22 potentially useful eHealth usability evaluation methods, and 11 not recommended eHealth usability evaluation methods were prioritized. The three most frequently recommended eHealth usability evaluation methods were remote user testing, expert review, and the rapid iterative testing

Supplementary Information The online version contains supplementary material available at https://doi.org/10.1007/978-3-658-44434-1_3.

I. Sinabell, *Agile eHealth Usability Evaluation*,
https://doi.org/10.1007/978-3-658-44434-1_3

and evaluation method. Eleven usability evaluation methods, such as retrospective testing, were not recommended for use in rapid eHealth usability evaluations. The comprehensive and evidence-based prioritization of eHealth usability evaluation methods supports faster usability evaluations and thus contributes to the ease-of-use of emerging eHealth systems.

Paper II:
Sinabell, I., & Ammenwerth, E. (2022). Challenges and Recommendations for eHealth Usability Evaluation with Elderly: Systematic Review and Case Study. *Universal Access in the Information Society*. https://doi.org/10.1007/s10209-022-00949-w

An explorative case study supplemented with a systematic literature review was conducted to learn about challenges in applying eHealth usability evaluation methods with elderly users and how these challenges can be overcome. Three established eHealth usability evaluation methods were chosen to evaluate the eHealth invention: (1) co-discovery evaluation, (2) cooperative usability testing, and (3) remote user testing combined with think aloud. The results showed that remote user testing combined with think aloud could successfully be applied to evaluate the eHealth invention with elderly users. Co-discovery evaluation and cooperative usability testing were not suitable to accomplish an agile eHealth usability evaluation with elderly users. The results showed that not all established eHealth usability evaluation methods are suitable for elderly users. Based on the case study and systematic literature review, 24 recommendations were developed on how to address challenges prior to, during, and after carrying out an eHealth usability evaluation. The recommendations were related to the stages of the software life cycle (requirements engineering, design, and evaluation) and age-related barriers (cognition, motivation, perception, and physical abilities).

Research Report:
Sinabell, I., & Ammenwerth, E. (2022). ToUsE: Toolbox for eHealth Usability Evaluations. Austria, Hall in Tirol: UMIT TIROL.

ToUsE ("Toolbox for eHealth Usability Evaluations") consists of rapidly applicable and potentially useful eHealth usability evaluation methods suitable for implementing eHealth usability evaluations, extended with further information, such as on the strengths and weaknesses of each eHealth usability evaluation method. ToUsE includes an overview of 43 eHealth usability evaluation methods related to their type (expert-based and user-based). The toolbox offers method cards consisting of a method description, including references of each eHealth usability evaluation method. The described eHealth usability evaluation methods are supplemented with information on aspects such as strengths, weaknesses, similarities, and/or relationships with other eHealth usability evaluation methods. ToUsE supports the selection of a rapidly applicable or potentially useful eHealth usability evaluation method to evaluate eHealth systems.

3.1 Research Phase 1: Identification and Expert Validation of 43 Rapidly Deployable and Potentially Useful eHealth Usability Evaluation Methods

RQ 1a: Which eHealth usability evaluation methods are easily applicable, rapidly deployable, and useful to support faster eHealth usability evaluations?

The findings of RQ 1a were published in *Applied Clinical Informatics*, an international journal published by Thieme.
Sinabell, I., & Ammenwerth, E. (2022). Agile, Easily Applicable, and Useful eHealth Usability Evaluations: Systematic Review and Expert-Validation. *Applied Clinical Informatics*, 13(01), 67–79. https://doi.org/10.1055/s-0041-1740919

Step 1:
After conducting the systematic literature review, a list of 29 eHealth usability evaluation methods consisting of easily applicable, rapidly deployable, and useful eHealth usability evaluation methods to support faster eHealth usability evaluations was identified (Figure 3.1). From these 29 identified eHealth usability evaluation methods, more were found to be user-based (e.g., remote user testing) than expert-based methods (e.g., consistency inspection). Overall, few of the eHealth usability evaluation methods found in the search correspond to the idea of discount usability engineering (micro tests and heuristic evaluation) (2/29). Most of the identified

eHealth usability evaluation methods represent traditional, established eHealth usability evaluation methods that were used in the literature to accomplish fast eHealth usability evaluations, such as heuristic evaluation, feature inspection, or questionnaires.

The minority of the identified eHealth usability evaluation methods, such as informal cognitive walkthrough or retrospective peer discovery, had evolved from traditional eHealth usability evaluation methods. Some of the identified eHealth usability evaluation methods can serve as an individual eHealth usability evaluation method (e.g., feature inspection), peer eHealth usability evaluation method (e.g., co-discovery evaluation), or group eHealth usability evaluation method (e.g., focus group).

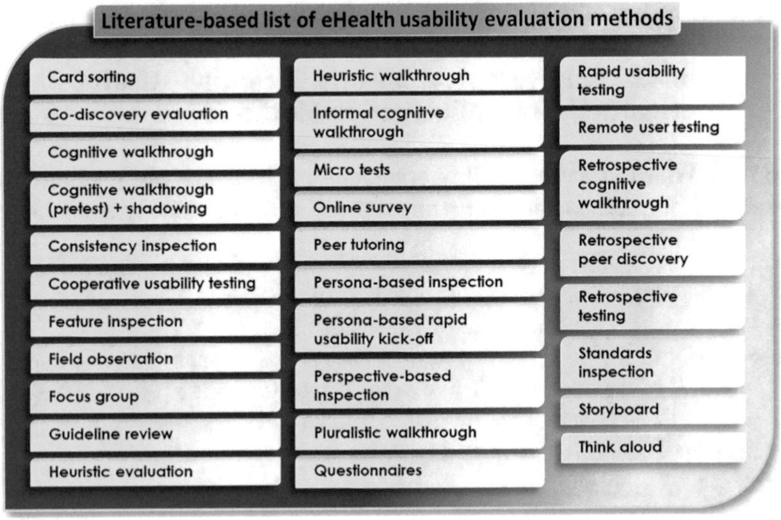

Figure 3.1 Literature-based list of agile eHealth usability evaluation methods modified from paper I (Sinabell & Ammenwerth, 2022). eHealth usability evaluation methods are arranged alphabetically

RQ 1b: How can the identified eHealth usability evaluation methods be prioritized to achieve agile, easily applicable, and useful eHealth usability evaluations?

> The findings of RQ 1b were published in *Applied Clinical Informatics*, an international journal published by Thieme.
> **Sinabell, I.**, & Ammenwerth, E. (2022). Agile, Easily Applicable, and Useful eHealth Usability Evaluations: Systematic Review and Expert-Validation. *Applied Clinical Informatics*, 13(01), 67–79.
> https://doi.org/10.1055/s-0041-1740919

Step 2a:
Overall, 43 eHealth usability evaluation methods were prioritized into ten recommended eHealth usability evaluation methods, 22 potentially useful eHealth usability evaluation methods, and 11 not recommended eHealth usability evaluation methods. The potentially useful eHealth usability evaluation methods include all those eHealth usability evaluation methods that experts had not commented on in more detail, i.e., they were neither recommended nor not recommended. The recommended eHealth usability evaluation methods refer to eHealth usability evaluation methods that experts recommended for use in rapid deployment. The not recommended eHealth usability evaluation methods refer to eHealth usability evaluation methods that experts did not recommend for use in rapid deployments. To achieve the prioritization of the identified eHealth usability evaluation methods, a list of 29 eHealth usability evaluation methods originating from the systematic literature review was extracted. The prioritization of the eHealth usability evaluation methods was achieved by iterative prioritization and refinement of the eHealth usability evaluation methods.

The three most frequently recommended eHealth usability evaluation methods are remote user testing, expert review, and the rapid iterative testing and evaluation method. Potentially useful eHealth usability evaluation methods for evaluating the eHealth usability are neither recommended nor not recommended by the experts. Perspective inspection, consistency inspection, standards inspection, and formal usability inspection, which were neither recommended nor not recommended as suitable for agile eHealth usability evaluations, should be applied early in the software life cycle. Retrospective usability testing is the most frequently recommended eHealth usability evaluation method, followed by focus groups, unmoderated usability testing, and questionnaires. The experts suggested the combination of think aloud and questionnaires as a way to supplement insufficiently meaningful qualitative results with quantitative results. Nevertheless, the experts do not recommend questionnaires to be used for agile eHealth usability evaluations, as a larger number of prospective users is required in order to obtain reliable quantitative results.

A detailed description of the iterative refinement and reasons why eHealth usability evaluation methods are recommended, potentially useful, or not recommended by experts can be found in the aforementioned paper provided in the ESM.

RQ 2: Which specific aspects for agile, easily applicable, and useful eHealth usability evaluations can be considered to evaluate eHealth inventions quickly?

The findings of RQ 2 are a secondary outcome of this work.

Step 2b:
The findings of RQ 1a and RQ 1b were supplemented with findings from two additional paper-based expert interviews. Based on these supplementary findings, a checklist for implementing an agile eHealth usability evaluation was formulated based on the experts' advice. Overall, the checklist comprises five categories: data collection, location and number of test sessions, prospective user group, test participants, and test performance (Figure 3.2). The checklist is intended for consideration by software developers or medical informaticians when conducting an agile eHealth usability evaluation.

Category one relates to the expert advice that data for agile eHealth usability evaluation should be collected qualitatively rather than quantitatively. The rationale is that "qualitative results are more meaningful than quantitative results". Experts take the view that "evaluating eHealth usability quantitatively is not purposeful because, for instance, questionnaire scoring systems usually do not meet the requirements for improving eHealth usability".

Category two includes the expert advice to test more often with fewer test participants because "during an ongoing agile development, only rapid testing should be applied". Secondly, the agile eHealth usability evaluation should be carried out in a real-world environment because experts share the view that "an on-site eHealth usability evaluation identifies the most revealing usability issues, which is useful in terms of agile eHealth usability evaluation".

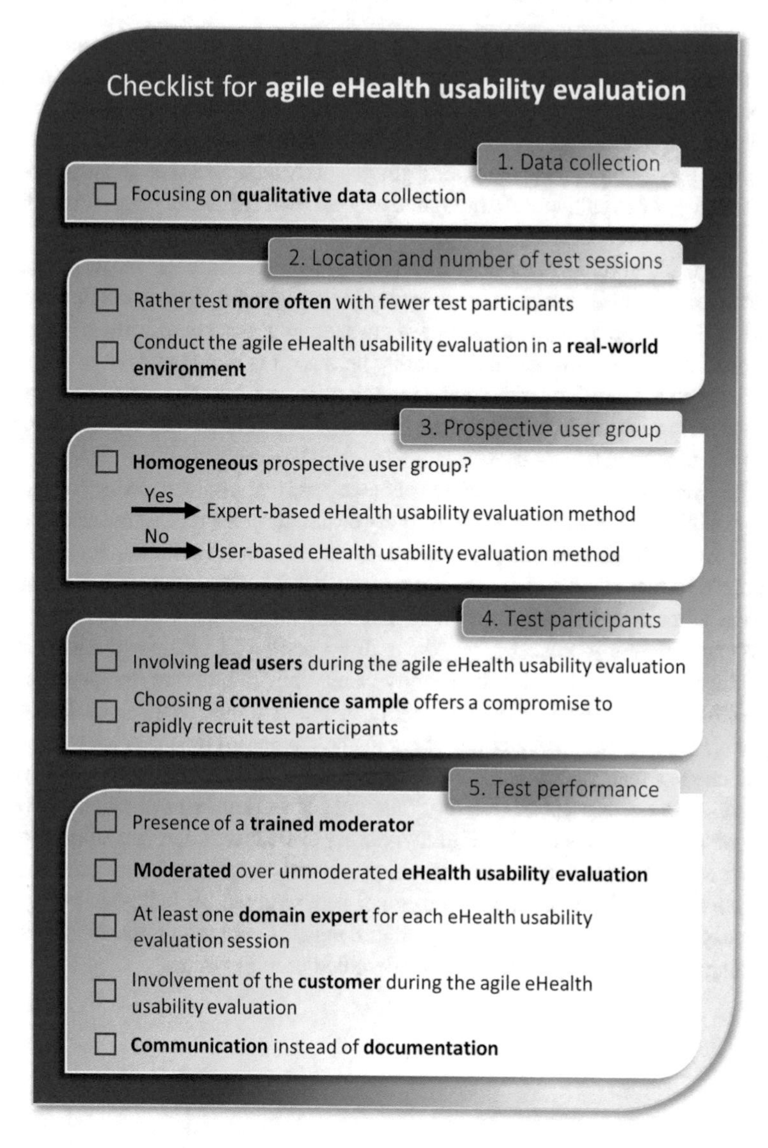

Figure 3.2 Checklist for implementing an agile eHealth usability evaluation. In the context of this book, the term lead user refers to users who are foremost in their field of expertise (Price et al., 2015). The term convenience sample refers to a sample of test participants "who are easily accessible in a clinical setting" (Setia, 2016). The categories of the checklist are arranged alphabetically

Category three refers to the advice to distinguish whether to apply an expert-based eHealth usability evaluation or user-based eHealth usability evaluation. The experts hold the view that "with user-based eHealth usability evaluations, the researcher can put him- or herself in the shoes of different prospective user groups, in the sense of nonhomogeneous user groups, and empathize with them". "Just as with the use of cognitive walkthrough, try to put yourself in the perspective of the prospective users".

Category four refers to the advice to involve lead users because "the involvement of lead users confirms whether we are going in the right direction or whether we got stuck". Furthermore, choosing a convenient sample of test participants accelerates the process of evaluating eHealth usability because it is important "not to waste so much time on recruiting representative test participants, but rely on people who are active in the work environment and are more readily available. This is often a viable compromise for obtaining useful results for eHealth usability".

Category five comprises the advice of the experts that a trained moderator should be present during an agile eHealth usability evaluation, as a moderated eHealth usability evaluation is preferable to an unmoderated eHealth usability evaluation. The rationale is that experts share the agreement that "during unmoderated usability testing, a lot of information gets lost" because "no questions can be asked, because no evaluator is present during the usability testing". Domain expertise should be integrated into agile eHealth usability evaluations in order not to "develop the eHealth invention past the medical users". Customers should be involved during an agile eHealth usability evaluation "to avoid and counteract time loss". The experts hold the view that "once you understand the concept of usability, you need to work according to the concept of usability. That means working efficiently, effectively, and satisfactorily". In an agile eHealth usability evaluation, communication should take precedence over documentation because experts believe that "When an eHealth invention is developed iteratively, customer needs should be addressed promptly. This can be achieved by immediately addressing customer requirements instead of creating detailed documentation of the identified usability issues".

3.2 Research Phase 2: ToUsE—Toolbox for eHealth Usability Evaluations

RQ 3: What could a toolbox for implementing agile, easily applicable, and useful ehealth usability evaluations look like?

The findings of RQ 3 were published in the following research report:
Sinabell, I., & Ammenwerth, E. (2022). ToUsE: Toolbox for eHealth Usability Evaluations. Austria, Hall in Tirol: UMIT TIROL.

Step 3:
Based on the findings from research phase 1, a toolbox called ToUsE was developed that comprises all 43 identified eHealth usability evaluation methods. ToUsE ("Toolbox for eHealth Usability Evaluations") consists of eHealth usability evaluation methods that are rapidly applicable and potentially useful for implementing fast eHealth usability evaluations.

All eHealth usability evaluation methods are listed on method cards with method descriptions. The method cards also contain references to similar or related eHealth usability evaluation methods, if any. ToUsE includes supplementary information on aspects such as strengths, weaknesses, similarities, and/or relationships with other eHealth usability evaluation methods described in ToUsE. The emphasis was consistently placed on the applicability of the described eHealth usability evaluation method for the context of eHealth, where available.

The eHealth usability evaluation methods in ToUsE were included regardless of expert prioritization because the focus was on eHealth systems as such and not on the applications in detail (e.g., mHealth systems or web-based eHealth systems). Thus, the not recommended eHealth usability evaluation methods were not excluded as not useful for performing fast eHealth usability evaluations. ToUsE supports the selection of a rapidly applicable or potentially useful eHealth usability evaluation method for evaluating eHealth inventions. The toolbox is intended for use by software developers, medical informaticians, usability experts, or medical professionals who plan to perform an eHealth usability evaluation.

The majority of eHealth usability evaluation methods in ToUsE are user-based (27/43), while the minority are expert-based (13/43) and a few are user-based and expert-based (4/43). An overview of all eHealth usability evaluation methods related to their type (expert-based and user-based) is included in ToUsE.

The overview of the expert-based eHealth usability evaluation methods and user-based eHealth usability evaluation methods included in ToUsE can be found in the aforementioned research report. A detailed description of the eHealth usability evaluation methods can be found in the research report as well, which is attached to the ESM.

3.3 Research Phase 3: Remote User Testing Combined With Think Aloud for Agile eHealth Usability Evaluation With Elderly Users

RQ 4: Which eHealth usability evaluation methods are suitable for evaluating eHealth interventions with elderly users?

The findings of RQ 4 were published in *Universal Access in the Information Society*, an international journal published by Springer.
Sinabell, I., & Ammenwerth, E. (2022). Challenges and Recommendations for eHealth Usability Evaluation with Elderly: Systematic Review and Case Study. *Universal Access in the Information Society*.
https://doi.org/10.1007/s10209-022-00949-w

Step 4:
All elderly participants were invited to complete the co-discovery evaluation, co-operative usability testing, and remote user testing combined with think aloud. The elderly users declined to participate in co-discovery evaluation as well as cooperative usability testing for a number of reasons. They preferred conducting tasks on their own and had concerns about the usefulness of video recordings. Elderly participants also expressed privacy concerns despite accurate explanations that personal information would be deleted after completing the eHealth usability evaluation.

The results of the case study were supplemented by the findings from the systematic literature review. These findings revealed that user-based eHealth usability evaluation methods are most commonly applied to conduct eHealth usability evaluations with elderly users.

In total, 20 papers that report on eHealth usability evaluation methods suitable for evaluating eHealth interventions with elderly users were analyzed.

The majority of the studies involved user-based eHealth usability evaluation methods, such as questionnaires, think aloud, or usability testing. Most of the studies used the eHealth usability evaluation method of questionnaires as a single or additional eHealth usability evaluation method. Of those studies applying questionnaires, the system usability scale was the most frequently used questionnaire. In the minority of studies, expert-based eHealth usability evaluation methods such as heuristic evaluation and cognitive walkthrough were employed to evaluate the usability of eHealth inventions. A retrospective eHealth usability evaluation was implemented as an additional eHealth usability evaluation method in two studies.

The studies focused on several eHealth systems that were evaluated, e.g., for health prevention, self-care, activity tracking and patient monitoring, and emergency care. Some investigated two eHealth systems, e.g., mHealth system combined with a web-based eHealth system.

A detailed description of the eHealth usability evaluation methods applied to elderly participants are shown in the aforementioned paper provided in the ESM. The results of the systematic literature review can be found in the aforementioned paper as well.

RQ 5: Is it feasible to conduct an agile eHealth usability evaluation with elderly users in a real-world environment?

The findings of RQ 5 were published in *Universal Access in the Information Society*, an international journal published by Springer.
Sinabell, I., & Ammenwerth, E. (2022). Challenges and Recommendations for eHealth Usability Evaluation with Elderly: Systematic Review and Case Study. *Universal Access in the Information Society*.
https://doi.org/10.1007/s10209-022-00949-w

Step 5:
Based on the findings of the case study, implementing an agile eHealth usability evaluation in a real-world environment proved to be feasible with elderly participants. The rationale is that the elderly participants felt comfortable in their familiar environment and were open-minded about using a new technology.

The findings from the case study reveal that remote user testing combined with think aloud could be successfully applied to evaluate eHealth systems with elderly users. However, not all established eHealth usability evaluation methods are suitable for conducting agile eHealth usability evaluations with elderly users. For example, the findings demonstrate that cooperative usability testing and co-discovery evaluation were not appropriate to evaluate eHealth systems with elderly participants. There are other eHealth usability evaluation methods that are not suitable for elderly users in their current version. These include co-discovery evaluation, cooperative usability testing, retrospective eHealth usability evaluation methods, and think aloud. Some of these eHealth usability evaluation methods have the potential to be more useful if they are evolved and adapted to meet the challenges of conducting eHealth usability evaluations with elderly participants.

Implementing the eHealth usability evaluation in a real-world environment was useful because assisting the elderly users with the steps of the eHealth usability evaluation was necessary (e.g., verbalizing their thoughts, comprehension of tasks). No video or audio recordings were made, as the elderly participants declined such recordings for privacy reasons. A variety of notable differences were observed in the way the elderly users interacted with the web-based eHealth invention. Some elderly participants had difficulties saving the prescription locally on their computer. This was not due to poor usability of the web-based eHealth invention but occurred because the elderly users were not sufficiently familiar with their own computer. Difficulties in task completion were noted due to motor impairments necessary to accomplish tasks, such as difficulties to enter data in the required fields of the web-based eHealth intervention during eHealth usability evaluation.

As a secondary outcome of this book, usability issues such as usability problems concerning the readability of the font size and textual or visual appearance were identified.

A detailed description of the findings from the case study and the identified usability issues can be found in the aforementioned paper provided in the ESM.

3.4 Research Phase 4: Twenty-four Recommendations for Addressing Challenges Prior to, During, and After Carrying Out the eHealth Usability Evaluation With Elderly Users

RQ 6a: What challenges may arise when conducting an agile, easily applicable, and useful eHealth usability evaluation with elderly users?

The findings of RQ 6a were published in *Universal Access in the Information Society*, an international journal published by Springer.
Sinabell, I., & Ammenwerth, E. (2022). Challenges and Recommendations for eHealth Usability Evaluation with Elderly: Systematic Review and Case Study. *Universal Access in the Information Society*.
https://doi.org/10.1007/s10209-022-00949-w

Step 6:
The results of the systematic literature review identified challenges due to motivational, cognitive, or physical decline of elderly users. These included challenges due to cognition resulted in limited verbalization skills, reduced attention, or speed of comprehension.

In total, six papers that report on the challenges of conducting eHealth usability evaluations with elderly or disabled users were analyzed. Challenges for elderly users were identified from four papers. For example, elderly users were reported to have difficulties with the execution of tasks due to motor impairments, short concentration spans, or forgotten intermediate steps, which made it necessary to repeat specific tasks. One study reported challenges due to motivation, for example because elderly users learn a new system more easily when they are supported by a mentor. One study also reported perceptual problems, for example that elderly users had difficulties in recognizing the font size of a text required for completing tasks during the eHealth usability evaluation. Similar challenges were identified for disabled people. For example, some had difficulties in verbalizing their thoughts or in interacting with the eHealth system due to motor impairments.

The case study experience provided evidence that elderly users had difficulties to express their thoughts aloud during the eHealth usability evaluation because of their rapidly decreasing attention. Elderly users were therefore encouraged to speak their thoughts aloud and were asked comprehension questions during the eHealth usability evaluation. Short tasks were used because the elderly participants tired quickly. In addition, the tasks were read aloud because this increased the elderly participants' attention and speed of understanding as to which tasks they should be working on. During the eHealth usability evaluation, motivational problems of the elderly users were noted, which explains why the elderly participants quickly lost their focus on tasks.

A detailed description of these findings can be found in the aforementioned paper provided in the ESM.

RQ 6b: How can the challenges that arise prior to, during, and after carrying out the eHealth usability evaluation conducted with elderly users be addressed and resolved?

The findings of RQ 6b were published in *Universal Access in the Information Society*, an international journal published by Springer.
Sinabell, I., & Ammenwerth, E. (2022). Challenges and Recommendations for eHealth Usability Evaluation with Elderly: Systematic Review and Case Study. *Universal Access in the Information Society*.
https://doi.org/10.1007/s10209-022-00949-w

Step 6:
From the case study and the systematic literature review, 24 recommendations were derived to address challenges prior to, during, and after the eHealth usability evaluation with elderly users. Overall, the recommendations were equally derived from the findings of the case study and the systematic review. Some of these case study-derived recommendations were also supported by the literature.

The challenges were categorized according to the stages of the software life cycle (requirements engineering, design, and evaluation), taking into account the stage following the eHealth usability evaluation. Two recommendations were formulated for the requirements engineering stage of the software life cycle, including the recommendation to apply the eHealth usability evaluation method using the personas technique or storyboards to reflect the needs of elderly users. Three recommendations were developed for the design stage of the software life cycle. One of these recommendations is to conduct an initial expert-based eHealth usability evaluation to identify major usability issues. Conducting a pilot study with elderly users prior to extensive eHealth usability evaluation is also recommended, as elderly users should ideally be involved at the design stage of the software life cycle. In total, 16 recommendations were derived for the evaluation stage of the software life cycle, such as to encourage elderly users to speak their thoughts aloud, which originates from the case study and was also supported by the literature. Another recommendation is to allow sufficient time for preparation and preliminary discussion of the eHealth usability evaluation with elderly users. For the stage following the eHealth usability evaluation, three recommendations were formulated. One of them includes the suggestion to provide continued support for the engagement of the elderly users.

The specific recommendations for the evaluation stage of the software life cycle were associated with age-related barriers (cognition, motivation, perception, and physical abilities). Of the 16 recommendations formulated for the evaluation stage of the software life cycle, most addressed motivation, while the second largest number of recommendations was associated with cognition, and the third largest number of recommendations dealt with physical abilities.

All 24 formulated recommendations that were derived from the case study and the systematic literature review can be found in the aforementioned paper provided in the ESM.

4 Discussion

The main research question of this book was as follows: Which usability evaluation approaches are appropriate and deployable to rapidly evaluate patient-centered eHealth inventions being iteratively developed in health care, and is it feasible to conduct an easily applicable eHealth usability evaluation as well as counter and tackle challenges that might arise during the agile eHealth usability evaluation with elderly users? This book aimed to examine the following research questions: (RQ 1a) Which eHealth usability evaluation methods are easily applicable, rapidly deployable, and useful to support faster eHealth usability evaluations? (RQ 1b) How can the identified eHealth usability evaluation methods be prioritized to achieve agile, easily applicable, and useful eHealth usability evaluations? (RQ 2) Which specific aspects for agile, easily applicable, and useful eHealth usability evaluations can be considered to evaluate eHealth inventions quickly? (RQ 3) What could a toolbox for implementing agile, easily applicable, and useful eHealth usability evaluations look like? (RQ 4) Which eHealth usability evaluation methods are suitable for evaluating eHealth interventions with elderly users? (RQ 5) Is it feasible to conduct an agile eHealth usability evaluation with elderly users in a real-world environment? (RQ 6a) What challenges may arise when conducting an agile, easily applicable, and useful eHealth usability evaluation with elderly users? (RQ 6b) How can the challenges that arise prior to, during, and after carrying out the eHealth usability evaluation conducted with elderly users be addressed and resolved?

4.1 Answering the Research Questions

The main research question was: *Which usability evaluation approaches are appropriate and deployable to rapidly evaluate patient-centered eHealth inventions being iteratively developed in health care, and is it feasible to conduct an easily*

I. Sinabell, *Agile eHealth Usability Evaluation*,
https://doi.org/10.1007/978-3-658-44434-1_4

applicable eHealth usability evaluation as well as counter and tackle challenges that might arise during the agile eHealth usability evaluation with elderly users?

The results of the triangulation study showed that the implementation of an agile, easily applicable, and useful eHealth usability evaluation with prospective users such as elderly patients proved to be feasible. Overall, 43 eHealth usability evaluation methods were identified as suitable for agile, easily applicable, and useful eHealth usability evaluations. Of these, ten were recommended by the experts based on their usefulness for rapid eHealth usability evaluations. The three most frequently recommended eHealth usability evaluation methods, regardless of their application to a prospective user group, were remote user testing, expert review, and rapid iterative test and evaluation method. However, the results showed that not all established eHealth usability evaluation methods are applicable with elderly users. Remote user testing combined with think aloud could successfully be applied to evaluate the eHealth invention with elderly users. The recommendations formulated for conducting an agile eHealth usability evaluation with elderly users illustrate the need to address the challenges of this age group because eHealth inventions should be tested with users who represent the intended prospective user group (Wildenbos, 2019a). The research results showed that involving elderly users early in the software life cycle is essential for the evolvement of eHealth inventions, which is in keeping with findings by Duque et al. (2019). The identified challenges in conducting an agile eHealth usability evaluation with elderly users reflected the way society is changing with and through information and communication technologies (Utesheva & Boell, 2016). Recent research has shown that health-related weaknesses often discourage elderly patients from using eHealth inventions (Airola, 2021). Conducting the explorative case study with the elderly participants revealed that health-related weaknesses due to age are not a barrier to their use. These results are consistent with Jakkaew & Hongthong (2017), who state that elderly people are open to adopting new technologies. This is crucial, as our society is changing to an information society (Leahy & Dolan, 2009).

Research question 1 was sub-divided into two separate questions (RQ1a and RQ1b). Research question 1a was: *Which eHealth usability evaluation methods are easily applicable, rapidly deployable, and useful to support faster eHealth usability evaluations?* The results showed that a variety of eHealth usability evaluation methods exist that are easily applicable, rapidly deployable, and useful to support faster eHealth usability evaluations. Twenty-nine such methods were identified from the literature. Several approaches to usability testing (cooperative usability testing, micro tests, or rapid usability testing) were identified as useful for eHealth usability evaluations that can be carried out quickly. The literature confirms the use of user-based eHealth usability evaluations for rapid deployments because, as can be seen from the example of usability testing, prospective users from the clinical environment (e.g., medical staff) can perform representative tasks to rapidly evaluate the usability of eHealth inventions that are employed in clinical practice (Kushniruk & Borycki, 2006). Nearly all of the identified approaches to eHealth usability evaluation involve established eHealth usability evaluation methods (e.g., heuristic walkthrough) published in the 2000s. There is a clear need for a variety of appropriate eHealth usability evaluation methods that are easily applicable, rapidly deployable, and useful to support faster eHealth usability evaluations, as many different types of eHealth inventions have been developed, deployed, and studied in practice over the past several years (Kip et al., 2022).

Research question 1b was: *How can the identified eHealth usability evaluation methods be prioritized to achieve agile, easily applicable, and useful eHealth usability evaluations?* In total, 43 eHealth usability evaluation methods were identified that are applicable to evaluating patient-centered eHealth inventions. Recent research revealed eHealth usability evaluation methods that are deployable to patient-centered eHealth inventions, such as electronic health patient activation (Haggerty et al., 2021), public health websites (Momenipour et al., 2021), or mHealth systems for the elderly (Wang et al., 2022). However, these methods have not been used to support faster eHealth usability evaluations. The identified eHealth usability evaluation methods were prioritized into ten recommended eHealth usability evaluation methods, 22 potentially useful eHealth usability evaluation methods, and 11 not recommended eHealth usability evaluation methods. For example, think aloud is suitable for iterative software development (Roberts & Fels, 2006) and can be applied as an iterative eHealth usability evaluation method to identify initial usability issues (Holzinger et al., 2012). This is confirmed by the fact that experts recommend the application of think aloud complementary to remote user testing for rapid eHealth usability evaluation. Approaches to heuristic evaluation are known as fast for the purpose of identifying usability issues (Coziahr et al., 2022). Inspection eHealth usability evaluation should be used complementary to usability testing and is regarded as fundamental to the prospective user group and the context of use (Silva et al., 2020). Knowledge of the context in which eHealth systems are to be evaluated is essential (Kuziemsky & Kushniruk, 2014) because it can impact the choice of a feasible combination of eHealth usability evaluation methods. Regardless of the context of eHealth usability evaluation, experts suggest combining qualitative and quantitative eHealth usability evaluation methods to enhance insufficiently informative qualitative results with quantitative results.

Research question 2 was: *Which specific aspects for agile, easily applicable, and useful eHealth usability evaluations can be considered to evaluate eHealth inventions quickly?* The results showed that a checklist with five categories (test cycles, test performance, data collection, prospective user group, and test user) was developed that can be applied prior to the implementation of an agile eHealth usability evaluation. Evaluating eHealth usability can cover the entire software life cycle (Lau & Kuziemsky, 2016), which underscores the need to consider the category of testing cycles. eHealth inventions "are designed to inform about, prevent, diagnose, treat, or monitor health conditions" (Broekhuis et al., 2021). This requires prospective users to, for instance, understand the health information provided by the system (Broekhuis et al., 2021). To enable prospective users to achieve this understanding and given the high complexity of eHealth systems (Liveri et al., 2015), such users must be involved in the agile development in terms of user-centered design from an early stage. Integrating user-centered design into the agile development process improves the quality and usability of the product (Hussain et al., 2009b). The author's view is supported by Lau & Kuziemsky (2016), who state that the scope of an eHealth usability evaluation may include the entire software life cycle. Formative usability testing precedes summative usability testing, which confirms the view of Broekhuis et al. (2021) that a "thorough understanding of eHealth usability supports formative evaluation methods". Testing more often with fewer test participants may require more highly developed technical and collaborative competencies of the moderator and the test participants (Wozney et al., 2016). The inclusion of a moderator in an agile eHealth usability evaluation supports the approach of eHealth usability evaluation methods such as the pluralistic walkthrough, in which a moderator introduces the system to be evaluated to test participants and invites them to discuss usability issues (Rödle et al., 2019). In addition to a user-based agile eHealth usability evaluation, an expert-based agile eHealth usability evaluation can also be performed. The latter focuses on the integration of a domain expertise. Dual-domain experts find a higher number of usability problems (Nielsen, 1992), which supports the need to involve at least one domain expert during an expert-based eHealth usability evaluation.

Research question 3 was: *What could a toolbox for implementing agile, easily applicable, and useful eHealth usability evaluations look like?* The ToUsE toolbox consists of method cards of all the eHealth usability evaluation methods identified in RQ 2. The described eHealth usability evaluation methods are supplemented by information on strengths, weaknesses, similarities, and/or relationships with other eHealth usability evaluation methods in ToUsE. These details provide an added value compared to existing toolboxes, such as the Electronic Health Record Usability Toolkit (Johnson et al., 2011), the Usability Body of Knowledge (UXPA, 2010), or the Usability Evaluation Toolbox of the University of Minnesota Duluth (UMD, 2022), which do not cover this information. One toolbox specifically addresses the evaluation of web-based eHealth inventions that includes different questionnaires (n=9) suitable for eHealth usability evaluation (Thielsch & Salaschek, 2020). ToUsE offers a much larger number of eHealth usability evaluation methods (n=43) that are rapidly applicable and potentially useful for implementing fast eHealth usability evaluations. Comparison research has been done into various eHealth usability evaluation methods; however, these mostly relate to user-based methods that are most frequently associated with eHealth usability, such as questionnaires (Broekhuis et al., 2019a; Maramba et al., 2019). ToUsE includes both user-based and expert-based eHealth usability evaluation methods. The authors Pohl & Scholz (2014) argue that a comparison of different usability evaluation methods is helpful for assessing their individual strengths and weaknesses. Although this suggestion was made for the field of information visualization, usability evaluation methods have strengths and weaknesses regardless of their application. The eHealth usability evaluation methods described in ToUsE were related to each other in terms of their strengths and weaknesses, if applicable. Contextual expertise, either single-domain expertise or dual-domain expertise, is required to select a suitable eHealth usability evaluation method. Georgsson et al. (2019) demonstrate that dual-domain experts (expertise in health care as well as in human-computer interaction) add an important additional dimension to eHealth usability evaluations. ToUsE addresses eHealth usability evaluation methods regardless of the prospective user group. This includes methods that can be used for the usability evaluation of digital health interventions directed at different user groups, such as children, adults, or the elderly (Ferrante et al., 2021).

Research question 4 was: *Which eHealth usability evaluation methods are suitable for evaluating eHealth interventions with elderly users?* The results from the case study showed that remote user testing combined with think aloud could be successfully applied to rapidly evaluate eHealth inventions with elderly users. This confirms that a combination of multiple eHealth usability evaluation methods is an effective way to evaluate usability (Georgsson, 2020). Some eHealth usability evaluation methods are recommended for use with the elderly (Wildenbos, 2019a). The results from the case study showed that not all established eHealth usability evaluation methods are applicable for an agile eHealth usability evaluation with elderly users. Cooperative usability testing and co-discovery evaluation were not suitable for evaluating eHealth inventions with elderly. Recent research has shown that questionnaires are widely employed to evaluate the usability of software designed for elderly users (Bastardo et al., 2022). However, questionnaires are considered unsuitable for evaluating usability for elderly people with cognitive impairments and need to be adapted for use with this age group (Yaddaden et al., 2019). The results of the case study confirmed the evolvement and adaptation of established eHealth usability evaluation methods, as remote user testing combined with think aloud had to be slightly adapted to be suitable for elderly users. For instance, no video or audio recordings were made due to the objections of elderly participants. Furthermore, the eHealth usability evaluations were conducted at the participants' homes and the software developers were connected via video conference due to the limited mobility of the elderly participants. The elderly users were open to the use of video-conferencing, which confirms that virtual group work is feasible with elderly users (Rietze et al., 2021). The results confirm the view of Roberts & Fels (2006) that think aloud should incorporate gestural and visual language. The implementation of the case study showed that think aloud is suitable for evaluating eHealth inventions with elderly users, provided these users are given strong motivation to speak their thoughts out loud.

Research question 5 was: *Is it feasible to conduct an agile eHealth usability evaluation with elderly users in a real-world environment?* Based on the results of the case study, implementing an agile eHealth usability evaluation in a real-world environment with elderly users proved to be feasible. The results also documented the feasibility of informal sessions of eHealth usability evaluation. These results are in line with the findings of a study conducted by Razak et al. (2013) who limited the number of recording devices brought to elderly participants' homes in order to reduce anxiety and stress. With elderly participants, aspects such as the use of larger fonts and buttons in the software need to be considered (Silva et al., 2020). This was confirmed by the case study, as one identified usability issue involved the overly small font size of the user interface of the web-based eHealth intervention. This experience is also in line with Springett et al. (2021) who found that elderly users have difficulty with finger-based input during task performance. This was confirmed by the case study, as the elderly participants had difficulties entering numbers. The results confirm the experience that elderly users felt more comfortable when the researcher (author of the book) was present during the evaluation (Razak et al., 2013). Health-related difficulties are the most common barrier to deploying eHealth systems among elderly users (Airola, 2021). Health-related difficulties affect not only the use of eHealth systems, but also the implementation of eHealth usability evaluations, as the case study findings confirmed. For example, the elderly participants expressed serious privacy concerns, which was one reason why no audio and video recordings were made during the agile eHealth usability evaluation. This experience confirms the need to address user concerns regarding privacy, including due to the increasing number of eHealth inventions such as mHealth systems (Camacho et al., 2020).

Research question 6 was subdivided into two research questions (RQ6a and RQ6b). Research question 6a was: *What challenges may arise when conducting an agile, easily applicable, and useful eHealth usability evaluation with elderly users?*

Several challenges were observed during the agile eHealth usability evaluation. For example, the elderly participants had difficulties speaking their thoughts out loud because their attention quickly decreased during the evaluation. The results revealed challenges due to motivational, cognitive, or physical decline of elderly users. Recent research has identified challenges countered by elderly patients when evaluating the usability of eHealth inventions, including physical impairments, such as difficulty with finger typing, or verbalization (Fan et al., 2021; Springett et al., 2021). These challenges were confirmed during the course of the case study, in which issues due to physical barriers, such as difficulties with the finger-based input due to limited hand movement, or with verbalization were observed. The elderly participants had difficulties speaking their thoughts out loud during the agile eHealth usability evaluation because their attention decreased as the evaluation progressed. For this reason, the elderly participants were expressly invited to speak their thoughts out loud. This experience is also in line with Wildenbos (2019a), who found that the application of think aloud strongly depends on the cognitive abilities of elderly participants. An eHealth usability study conducted with elderly users by Isaković et al. (2016) revealed that elderly persons easily lose the thread during eHealth usability evaluations and that their concentration decreases. This experience was confirmed by the case study. When evaluating the usability of the eHealth invention, it became clear that the elderly users tired quickly, which is why only three short tasks were given. The experience of Springett et al. (2021) who emphasized the benefits of task cooperating for elderly users was not confirmed, as the elderly participants explicitly rejected the idea of joint task processing.

Research question 6b was: *How can the challenges that arise prior to, during, and after carrying out the eHealth usability evaluation conducted with elderly users be addressed and resolved?* From the findings of the case study and the systematic review, 24 recommendations were derived to address challenges prior to, during, and after eHealth usability evaluations with elderly users. Recommendations for proposed eHealth usability evaluations with elderly users can be found in the literature (Engelsma et al., 2021). However, these suggestions are aimed at mHealth design for elderly patients suffering from Alzheimer's disease and related types of dementia. These suggestions addressed age-related barriers, such as cognitive, physical, perceptual, or mental barriers. The 24 formulated recommendations were categorized according to the stages of the software life cycle (requirements engineering, design, and evaluation), taking into account the stage following the eHealth usability evaluation. In addition, the recommendations for the evaluation stage of the software life cycle were associated with age-related barriers (cognition, motivation, perception, and physical abilities). Overall, the formulated recommendations addressed ways to handle challenges prior to, during, and after carrying out the eHealth usability evaluation with elderly users. Considering the stages before and after the eHealth usability evaluation puts the eHealth usability evaluation in a holistic context. A holistic view also brings prospective users into the equation. The developed recommendations show different ways to improve eHealth usability for elderly users. Designing user interfaces for the disabled and the elderly is an important topic for human-computer interaction (Marcus, 2003). Universal usability emphasizes developing products that can be used by as many people in as many situations as possible (Vanderheiden, 2020). In the spirit of considering universal usability, challenges for implementing an eHealth usability evaluation with disabled users were identified despite the fact that they are not the intended user group of the web-based eHealth invention under study.

4.2 Limitations

A triangulative approach combining different qualitative methods was chosen to implement the book. Based on the initial systematic literature review of this study, more than 300 relevant papers were ultimately included. To avoid overlooking relevant papers, the snowball principle was applied, and additional relevant papers were identified using a manual search.

The iterative qualitative interviews with experts were conducted in the experts' native language. A broad selection of experts was considered, but most of the experts were affiliated with research and development. Qualitative content analysis was applied to analyze the transcripts of the expert interviews. In the course of the content analysis, eHealth usability evaluation methods were prioritized in terms of their usefulness. For this purpose, quantitative frequencies of the eHealth usability evaluation methods were counted. As the expert interviews were conducted iteratively, the results of one iteration may affect the subsequent iteration; however, saturation of the results was observed after the fourth iteration.

The specifics to facilitate the implementation of an agile eHealth usability evaluation represent a collection of expert knowledge. The specifics derived from the experts' statements have not been validated against each other. However, to support the specifics derived from the experts' statements, two additional written expert interviews were conducted with evolved questions taken from the semi-structured interview guide. Key expert quotes were included in the results section to justify the selection of specifics.

The ToUsE toolbox was developed to facilitate the selection of a rapidly applicable and potentially useful eHealth usability evaluation method. To enable a sound selection of an appropriate eHealth usability evaluation method, all identified methods (n=43) were included. The toolbox is intended for use by software developers, medical informaticians, usability experts, or medical professionals who plan to perform an eHealth usability evaluation. Of course, the application is facilitated by dual-domain experts with special knowledge in the two fields of usability and health care. The added value of ToUsE compared to existing toolboxes is that it includes extended information on strengths, weaknesses, similarities and/or relationships with other eHealth usability evaluation methods. This knowledge was derived from the findings of the initial systematic literature review, but it cannot be assumed that all information has been identified. An additional manual search was therefore conducted specifically for the development of ToUsE, in the course of which over 65 additional relevant papers were identified.

Participants suffering from chronic diseases were selected for the explorative case study. Over the course of the case study, elderly participants differed in how they wanted to be involved in the eHealth usability evaluation depending on their health status and the severity of age-related barriers. Different challenges to age-related barriers were identified, most of them related to cognition and perception. Saturation was achieved after implementing the eHealth usability evaluation with the fourth elderly participant due to repetition of the usability issues that were found. However, it cannot be assumed that all possible challenges were identified. All elderly participants graduated at least from high school; more than one-third had a university degree. It cannot be ruled out that the level of education affected the identified usability problems, but basic computer skills constituted an inclusion criterion for participation in the eHealth usability evaluation.

During the second systematic literature review to complement the findings from the case study, different search terms and MeSH terms were used. It cannot be assumed that all existing approaches to agile eHealth usability evaluation were found within the systematic literature review, although care was taken in the search to use keywords in different grammatical styles. To obtain reliable results, four different databases (ACM Digital Library, Cochrane Library, IEEE Xplore, and Medline) were used for searching and retrieving papers. Taking into consideration that some findings may have been missed in the gray literature, an additional manual search was conducted.

The recommendations derived from the case study and the systematic literature were related to the stage of the software life cycle in order to consider their relevance for software developers. The developed recommendations are limited to conducting agile eHealth usability evaluations with elderly participants, although the challenges were also considered from the related field of disabled users through the systematic literature search.

4.3 New Research Questions

Additional research is needed to consolidate the recommendations from the findings of the systematic literature review and case study in a similar context of eHealth inventions and other prospective user groups. To get closer to achieving universal usability, the recommendations should be validated with other prospective user groups, such as disabled users, and expanded specifically for them. This issue could be investigated by answering the following new research question: How can the recommendations be expanded to other prospective user groups to facilitate the implementation of universal eHealth usability?

Overall, the results have shown that not all established eHealth usability evaluation methods are suitable for conducting agile eHealth usability evaluations with elderly users. The evolvement of eHealth usability evaluation methods suitable for conducting eHealth usability evaluations with elderly users should be investigated further. The following new research question can be derived: How can eHealth usability evaluation methods suitable for agile eHealth usability evaluation be evolved to address challenges presented by elderly users?

Within the scope of this book, ToUsE was developed as a toolbox comprising different rapidly applicable and potentially useful eHealth usability evaluation methods. Contextual eHealth knowledge is required to select an appropriate eHealth usability evaluation method. Future research is also required to support the context-based selection of an appropriate eHealth usability evaluation method. This results in a new research question: What could a decision tree for context-based selection of an appropriate eHealth usability evaluation method from ToUsE look like?

Conclusion

5

This book aimed to (1) systematically identify and expert validate rapidly deployable eHealth usability evaluation methods to support faster eHealth usability evaluations, (2) develop a toolbox for agile eHealth usability evaluation, (3) examine the applicability of eHealth usability evaluation methods to elderly users and the feasibility of conducting an agile eHealth usability evaluation with elderly users, and (4) identify, dealing with, and overcome challenges that might be countered prior to, during, and after carrying out an eHealth usability evaluation with elderly users.

The absence of existing approaches on agile eHealth usability evaluation that are suitable for elderly users led to the identification of suitable eHealth usability evaluation methods applicable in this context, the selection of established eHealth usability evaluation methods to accomplish the eHealth usability evaluation with elderly users, and the development of 24 recommendations to address challenges prior to, during, and after carrying out an eHealth usability evaluation with elderly users. The application of established eHealth usability evaluation methods revealed that not all of these methods are helpful with elderly users. While none of the eHealth usability evaluation methods were suitable for elderly users on their own, the use of remote user testing combined with think aloud formed the foundation of the successful implementation of the explorative case study. An evolvement of established eHealth usability evaluation methods is necessary to consider the particular needs of elderly users due to age-related declines in terms of cognition, motivation, perception, and physical abilities.

The findings of the expert validation and systematic review of rapidly deployable eHealth usability evaluation methods published in paper I "Agile, Easily Applicable, and Useful eHealth Usability Evaluations: Systematic Review and Expert-Validation" support the implementation of an agile eHealth usability evaluation and contribute to enhancing the usability of eHealth systems. The paper shows that certain eHealth usability evaluation methods are recommended for agile, easily ap-

I. Sinabell, *Agile eHealth Usability Evaluation*,
https://doi.org/10.1007/978-3-658-44434-1_5

plicable, and useful eHealth usability evaluations. The three eHealth usability evaluation methods most frequently recommended by experts are remote user testing combined with think aloud, expert review, and the rapid iterative testing and evaluation method. Several eHealth usability evaluation methods, such as retrospective testing, focus groups, and unmoderated usability testing, were deemed less helpful for conducting an agile eHealth usability evaluation and were not recommended by the experts. Some of these approaches were theoretically conceived, which indicates the lack of practical implementation and the urgent need for an explorative investigation of such approaches. These findings regarding prioritized eHealth usability evaluation methods are pertinent to patient-centered eHealth systems as such, which contribute to meaningful improvements to the usability of eHealth systems and thus enable accessible health care for the aging population.

The recommendations for conducting an agile eHealth usability evaluation with elderly participants derived from the explorative case study findings, complemented by the results from the second systematic literature review published in paper II "Challenges and Recommendations for eHealth Usability Evaluation with Elderly Users: Systematic Review and Case Study", are intended to help improve the accessibility, acceptability, and usability of eHealth systems for elderly users. The majority of the formulated recommendations relate to challenges concerning cognition and motivation. It became apparent that one chosen eHealth usability evaluation method - remote user testing combined with think aloud - proved to be suitable for agile eHealth usability evaluations with elderly users. Co-discovery evaluation and cooperative usability testing, on the other hand, were not suitable for eHealth usability evaluation with elderly users. The findings of the paper showed that the eHealth usability evaluation methods applied were mostly user-based, which illustrates the necessity of involving elderly users in the iterative development of eHealth systems. These developed recommendations are confined to eHealth usability evaluations with elderly users, making future work in the evolvement of these recommendations necessary so they can become applicable to further prospective user groups such as the disabled.

The newly developed ToUsE toolbox published in the research report "ToUsE: Toolbox for eHealth Usability Evaluations" provides descriptions of eHealth usability evaluation methods that are complemented by strengths, weaknesses, similarities, and/or relationships with other eHealth usability evaluation methods, which support the selection of a suitable eHealth usability evaluation method. The toolbox provides a comprehensive outline of rapidly deployable and potentially useful eHealth usability evaluation methods for all prospective user groups of eHealth usability evaluations. Future research should focus on the context-based selection of appropriate eHealth usability evaluation methods from ToUsE.

The developed checklist, a secondary result of this book, offers some specifications for conducting an agile eHealth usability evaluation. The checklist is aimed at software developers, medical informaticians, and usability experts who plan to perform an agile eHealth usability evaluation. To make the checklist relevant for clinical practitioners, a validation in a clinical setting, e.g., municipal hospital, is needed. To fine-tune the specifications of the checklist and enhance its relevance for software developers, it would be useful to relate the specifications to the stages of the software life cycle. Considering the different pieces of advice mentioned in the checklist (e.g., gathering qualitative data over quantitative data) ensures a rapid way to improve the preparation of agile eHealth usability evaluations throughout the iterative development of eHealth systems.

The findings of this book facilitate the employment of agile eHealth usability evaluations as a step towards universal usability of eHealth inventions. This book contributed to the development of improved patient-centered eHealth systems tailored to elderly users. The recommendations of this book foster evidence-based fast and flexible evaluation of eHealth inventions that contribute to further innovations in health care.

List of Publications

Articles in international journals (peer-reviewed)

Sinabell, I., & Ammenwerth, E. (2022). Agile, Easily Applicable, and Useful eHealth Usability Evaluations: Systematic Review and Expert-Validation. *Applied Clinical Informatics*, 13(01), 67–79. https://doi.org/10.1055/s-0041-1740919

Sinabell, I., & Ammenwerth, E. (2022). Challenges and Recommendations for eHealth Usability Evaluation with Elderly: Systematic Review and Case Study. *Universal Access in the Information Society*. https://doi.org/10.1007/s10209-022-00949-w

Research Report

Sinabell, I., & Ammenwerth, E. (2022). ToUsE: Toolbox for eHealth Usability Evaluations. Austria, Hall in Tirol: UMIT TIROL.

Other contributions (ordered descending by year)

Sinabell, I. Agile Usability Engineering von eHealth-Anwendungen. Presentation at the 13th GMDS PhD Symposium, RWTH Aachen University, University Hospital Aachen, Germany, September 28, 2022.

Sinabell, I. Agile Usability Evaluation patientenzentrierter e-Health Technologien. Poster Presentation at the Vienna School of Methods in cooperation with the International EARLI SIG17+8 conference 2020 at the University of Vienna, Austria, September 07, 2020.

Sinabell, I. Patient-centered eHealth technologies: Agile Usability Evaluation on the example of elderly as prospective users. Presentation at the TTRN 2019 PhD course *Research methods in innovation processes for digital health technology*, Graduate School of Health Sciences, University of Southern Denmark, Odense, Denmark, August 12, 2019.

I. Sinabell, *Agile eHealth Usability Evaluation*, https://doi.org/10.1007/978-3-658-44434-1

Glossary

Table I Key terms and their definitions that apply in this book

Term	Definition
Agile	"Agile is the ability to create and respond to change" (Agile Alliance, 2022).
Agility	Agility can be briefly defined as "responsiveness to change" (Gren & Lenberg, 2020). In this book, agility is defined as the possibility to rapidly implement an agile, easily applicable, and useful eHealth usability evaluation necessary to obtain early user feedback to improve the usability of an eHealth system.
Agile eHealth usability evaluation	In this book, an agile eHealth usability evaluation is defined as the possibility to easily realize a usability evaluation that can be performed rapidly.
Agile development	"Agile development involves the iterative, step-by-step development of software" (Kushniruk & Borycki, 2015). An advantage of agile development is that changes to the software can be incorporated quickly and users can be integrated into the software development process at an early stage (Kushniruk & Borycki, 2015).
eHealth	eHealth is defined as "[…] an emerging field in the intersection of medical informatics, public health, and business, referring to health services and information delivered or enhanced through the internet and related technologies" (Eysenbach, 2001).
eHealth evaluation	eHealth evaluation is applied to evaluate the usability of eHealth systems (Lau & Kuziemsky, 2016).

(Continued)

I. Sinabell, *Agile eHealth Usability Evaluation*,
https://doi.org/10.1007/978-3-658-44434-1

Table I (Continued)

Term	Definition
eHealth usability evaluation method	eHealth usability evaluation methods are applied to accomplish eHealth usability evaluations, which can be expert-based, user-based, or a combination of both expert- and user-based.
eHealth systems	eHealth systems include any information and communication technologies-based applications that can be applied in health care (Lau & Kuziemsky, 2016). In this book, a system is regarded as "a product or as the services it provides" (International Standardization Organization, 2018), which may include a variety of eHealth inventions (e.g., mHealth or web-based eHealth systems).
Elderly	Elderly can be regarded as persons with a chronological age equal to or older than 65 years (Orimo et al., 2006).
Information and communication technology	Information and communication technology can be used in a supportive manner in health care, for example, to monitor the health status of patients (Stara et al., 2021).
mHealth	mHealth (mobile health) systems enable a decentralized transfer of health data in real time to provide health care delivery (Isaković et al., 2016).
Representative sample of test users	A representative sample of test users refers to between five and ten test users with whose help most of the usability issues can be identified (Kushniruk & Borycki, 2006).
Usability	Usability is defined as the "extent to which a system, product or service can be used by specified users to achieve specified goals with effectiveness, efficiency and satisfaction in a specified context of use" (International Standardization Organization, 2018). For the completeness of the definition of usability, the definitions of effectively, efficiently, and satisfactorily are included here: effectiveness is defined as "accuracy and completeness with which users achieve specified goals"; efficiency is defined as "resources used in relation to the results achieved"; and satisfaction is defined as the "extent to which the user's physical, cognitive, and emotional responses that result from the use of a system, product or service meet the user's needs and expectations" (International Standardization Organization, 2018).

(Continued)

Table I (Continued)

Term	Definition
Usability engineering	"Usability engineering is the methodical way to achieve the feature of usability" (Heimgärtner, 2017). The feature of usability refers to the effectiveness, efficiency, and satisfaction of usability. Usability engineering aims at providing practical feedback to accomplish the development of software (Kushniruk & Patel, 2004).
Usability evaluation	Usability evaluation can be analytical (i.e., applying expert-based eHealth usability evaluation methods) or empirical (i.e., applying user-based eHealth usability evaluation methods) (Silva et al., 2020).
Usability testing	Usability testing is used to improve the usability of software, in the course of which representative end users of a system are observed performing tasks (Kushniruk & Borycki, 2006). Usability tests thus show how a future user interacts with a software, which can differ from the ideas of the software developers (Russ et al., 2010).
User	A user is defined as a "person who interacts with a system, product, or service" (International Standardization Organization, 2018).
User experience	A more recent concept on encompassing usability is user experience (Barnard et al., 2013).
Patient-centeredness	In this book, patient-centeredness refers not only to patient-centered care as such; it focuses more on patients and their health needs in order to improve health care services. For completeness of the definition, patient-centered care is defined "as health care that respects and responds to the preferences, needs, and values of the individual patients throughout all health care decisions" (Institute of Medicine, 2018).
User group	A user group is defined as a "subset of intended users who are differentiated from other intended users by characteristics of the users, tasks, or environments that could influence usability" (International Standardization Organization, 2018).

Bibliography

Agile Alliance. (2022). What is Agile? Retrieved November 9, 2022, from https://www.agilealliance.org/agile101/

Aguirre, R. R., Suarez, O., Fuentes, M., & Sanchez-Gonzalez, M. A. (2019). Electronic Health Record Implementation: A Review of Resources and Tools. *The Cureus Journal of Medical Science*, 11(9), Article e5649. https://doi.org/10.7759/cureus.5649

Airola, E. (2021). Learning and Use of eHealth Among Older Adults Living at Home in Rural and Nonrural Settings: Systematic Review. *Journal of Medical Internet Research*, 23(12), Article e23804. https://doi.org/10.2196/23804

Alajarmeh, N., Pontelli, E., & Son, T. (2011). From "Reading" Math to "Doing" Math: A New Direction in Non-visual Math Accessibility. In C. Stephanidis (Ed.) *Lecture Notes in Computer Science: Vol. 6768. Universal Access in Human-Computer Interaction. Applications and Services* (pp. 501–510). Springer. https://doi.org/10.1007/978-3-642-21657-2_54

Ammenwerth, E., Iller, C., & Mansmann, U. (2003). Can evaluation studies benefit from triangulation? A case study. *International Journal of Medical Informatics*, 70(2–3), 237–248. https://doi.org/10.1016/S1386-5056(03)00059-5

Arning, K., & Ziefle, M. (2009). Different Perspectives on Technology Acceptance: The Role of Technology Type and Age. In A. Holzinger & K. Miesenberger (Eds.), *Lecture Notes in Computer Science: Vol. 5889. HCI and Usability for e-Inclusion* (pp. 20–41). Springer. https://doi.org/10.1007/978-3-642-10308-7_2

Badawy, S. M., Cronin, R. M., Hankins, J., Crosby, L., DeBaun, M., Thompson, A. A., & Shah, N. (2018). Patient-Centered eHealth Interventions for Children, Adolescents, and Adults With Sickle Cell Disease: Systematic Review. *Journal of Medical Internet Research*, 20(7), Article e10940. https://doi.org/10.2196/10940

Barnard, Y., Bradley, M. D., Hodgson, F., & Lloyd, A. D. (2013). Learning to use new technologies by older adults: Perceived difficulties, experimentation behaviour and usability. *Computers in Human Behavior*, 29(4), 1715–1724. https://doi.org/10.1016/j.chb.2013.02.006

Bastardo, R., Pavão, J., & Pacheco Rocha, N. (2022). User-Centred Usability Evaluation of Embodied Communication Agents to Support Older Adults: A Scoping Review. In Á. Rocha, C. Ferrás, A. Méndez Porras, & E. Jimenez Delgado (Eds.), *Lecture Notes in Networks and Systems: Vol. 414. Information Technology and Systems* (pp. 509–518). Springer. https://doi.org/10.1007/978-3-030-96293-7_42

I. Sinabell, *Agile eHealth Usability Evaluation*,
https://doi.org/10.1007/978-3-658-44434-1

Bastien, J. M. C. (2010). Usability testing: a review of some methodological and technical aspects of the method. *International Journal of Medical Informatics*, 79(4), Article e18–e23. https://doi.org/10.1016/j.ijmedinf.2008.12.004

Beck, K., Beedle, M., van Bennekum, A., Cockburn, A., Cunningham, W., Fowler, M., Grenning, J., Highsmith, J., Hunt, A., Jeffries, R., Kern, J., Marick, B., Martin, R. C., Mellor, S., Schwaber, K., Sutherland, J., & Thomas, D. (2001). *Manifesto for Agile Software Development.* https://agilemanifesto.org/

Bekhet, A. K., & Zauszniewski, J. A. (2012). Methodological triangulation: an approach to understanding data. *Nurse Researcher*, 20(2), 40–3. https://doi.org/10.7748/nr2012.11.20.2.40.c9442

Berg, B., Knott, P., & Sandhaus, G. (2014). *Hybride Softwareentwicklung. Das Beste aus klassischen und agilen Methoden in einem Modell vereint.* Springer-Verlag. https://doi.org/10.1007/978-3-642-55064-5

Blankenhagel, K. J. (2019). Identifying Usability Challenges of eHealth Applications for People with Mental Disorders: Errors and Design Recommendations. In Association for Computing Machinery (Ed.), *PervasiveHealth 2019: Proceedings of the 13th International Conference on Pervasive Computing Technologies for Healthcare* (pp. 91–100). https://doi.org/10.1145/3329189.3329195

Bogner, A., Littig, B., & Menz, W. (2002). *Das Experteninterview. Theorie, Methode, Anwendung.* Springer Fachmedien. https://doi.org/10.1007/978-3-322-93270-9

Bogner, A., Littig, B., & Menz, W. (2014). *Interviews mit Experten: Eine praxisorientierte Einführung.* Springer Fachmedien. https://doi.org/10.1007/978-3-531-19416-5

Bonis, P. A. (2019). Clinical Decision Support Improves Decision-Making and Leads to Better Outcomes. *Acta Médica Portuguesa*, 32(10), 677. https://doi.org/10.20344/amp.12688

Borycki, E., Kushniruk, A., Nohr, C., Takeda, H., Kuwata, S., Carvalho, C., Bainbridge, M., & Kannry, J. (2013). Usability Methods for Ensuring Health Information Technology Safety: Evidence-Based Approaches. Contribution of the IMIA Working Group Health Informatics for Patient Safety. *Yearbook of Medical Informatics*, 22(1), 20–7. https://doi.org/10.1055/s-0038-1638828

Brandt-Pook, H., & Kollmeier, R. (2008). *Softwareentwicklung kompakt und verständlich. Wie Softwaresysteme entstehen.* Springer Vieweg. https://doi.org/10.1007/978-3-8348-9507-3

Broekhuis, M., van Velsen, L., & Hermens, H. (2019a). Assessing usability of eHealth technology: A comparison of usability benchmarking instruments. *International Journal of Medical Informatics*, 128, 24–31. https://doi.org/10.1016/j.ijmedinf.2019.05.001

Broekhuis, M., van Velsen, L., Peute, L., Halim, M., & Hermens, H. (2021). Conceptualizing Usability for the eHealth Context: Content Analysis of Usability Problems of eHealth Applications. *JMIR Formative Research*, 5(7), Article e18198. https://doi.org/10.2196/18198

Broekhuis, M., van Velsen, L., ter Stal, S., Weldink, J., & Tabak, M. (2019b). Why My Grandfather Finds Difficulty in using Ehealth: Differences in Usability Evaluations between Older Age Groups. In M. Ziefle, & Maciaszek, L. (Eds.), *ICT4AWE 2019: Proceedings of the 5th International Conference on Information and Communication Technologies for Ageing Well and e-Health* (pp. 48–57). Association for Computing Machinery. https://doi.org/10.5220/0007680800480057

Bundesministerium für Verkehr, Innovation und Technologie (BMVIT) (2019). Breitbandstrategie 2030. Österreichs Weg in die Gigabit-Gesellschaft. BMVIT.

Camacho, E., Hoffman, L., Lagan, S., Rodriguez-Villa, E., Rauseo-Ricupero, N., Wisniewski, H., Henson, P., & Torous, J. (2020). Camacho E, Hoffman L, Lagan S, et al. Technology Evaluation and Assessment Criteria for Health Apps (TEACH-Apps): Pilot Study. *Journal of Medical Internet Research*, 22(8), Article e18346. https://doi.org/10.2196/18346

Cao, Q., & Cheng, X. (2022). Characteristics of Interaction Design and Advantages of Network Teaching. In M. M. Soares, E. Rosenzweig, & A. Marcus (Eds.), *Lecture Notes in Computer Science: Vol. 13322. Design, User Experience, and Usability: Design for Emotion, Well-being and Health, Learning, and Culture* (pp. 256–264). Springer. https://doi.org/10.1007/978-3-031-05900-1_17

Cavichi de Freitas, R., Rodrigues, L. A., & Marques da Cunha, A. (2016). AGILUS: A Method for Integrating Usability Evaluations on Agile Software Development. *Proceedings of the 18th International Conference on Human-Computer Interaction, Toronto, ON, Canada*, 545–552. https://doi.org/10.1007/978-3-319-39510-4_50

Chiarini, G., Ray, P., Akter, S., Masella, C., & Ganz, A. (2013). mHealth Technologies for Chronic Diseases and Elders: A Systematic Review. *IEEE Journal on Selected Areas in Communications*, 31(9), 6–18. https://doi.org/10.1109/JSAC.2013.SUP.0513001

Coziahr, K., Stanley, L., Perez-Litwin, A., Lundberg, C., & Litwin, A. (2022). Designing a Digital Mental Health App for Opioid Use Disorder Using the UX Design Thinking Framework. In M. M. Soares, E. Rosenzweig, & A. Marcus (Eds.), *Lecture Notes in Computer Science: Vol. 13322. Design, User Experience, and Usability: Design for Emotion, Well-being and Health, Learning, and Culture* (pp. 107–129). Springer. https://doi.org/10.1007/978-3-031-05900-1_7

Cunha, A., Trigueiros, P., & Lemos, T. (2014). Reassuring the Elderly Regarding the Use of Mobile Devices for Mobility. In C. Stephanidis & M. Antona (Eds.), *Lecture Notes in Computer Science: Vol. 8515. Universal Access in Human-Computer Interaction. Aging and Assistive Environments* (pp. 46–57). Springer. https://doi.org/10.1007/978-3-319-07446-7_5

da Silva Santos, K., Ribeiro, M. C., Ulisses de Queiroga, D. E., Pereira da Silva, I. A., & Soares Ferreira, S. M. (2020). The use of multiple triangulations as a validation strategy in a qualitative study. *Ciencia & saude coletiva*, 25(2), 655–664. https://doi.org/10.1590/1413-81232020252.12302018

da Silva, T. S., Selbach Silveira, M., & Maurer, F. (2015). Usability Evaluation Practices within Agile Development. *Proceedings of the 48th Hawaii International Conference on System Science, HICSS 2015, Kauai, HI, USA*, 5133–5142. https://doi.org/10.1109/HICSS.2015.607

Dix, A. (2009). Human-Computer Interaction. In L. Liu & M. T. Özsu (Eds.), *Encyclopedia of Database Systems* (pp. 1327–1331). Springer. https://doi.org/10.1007/978-0-387-39940-9_192

Downey, L. L. (2007). Group usability testing: evolution in usability techniques. *Journal of Usability Studies*, 2(3), 133–144.

Duque, E., Fonseca, G., Vieira, H., Gontijo, G., & Ishitani, L. (2019). A systematic literature review on user centered design and participatory design with older people. In Association for Computing Machinery (Ed.), *IHC 2019: Proceedings of the 18th Brazilian Symposium on Human Factors in Computing Systems* (pp. 1–11). https://doi.org/10.1145/3357155.3358471

Engelsma, T., Jaspers, M. W. M., & Peute, L. W. (2021). Considerate mHealth design for older adults with Alzheimer's disease and related dementias (ADRD): A scoping review on usability barriers and design suggestions. *International Journal of Medical Informatics*, 152, 104494. https://doi.org/10.1016/j.ijmedinf.2021.104494

Eysenbach, G. (2001). What is e-health? *Journal of Medical Internet Research*, 3(2), e20. https://doi.org/10.2196/jmir.3.2.e20

Eysenbach, G. (2018). CONSORT-EHEALTH: Improving and Standardizing Evaluation Reports of Web-based and Mobile Health Interventions. *Journal of Medical Internet Research*, 13(4), Article e126. https://doi.org/10.2196/jmir.1923

Fan, M., Zhao, Q., & Tibdewal, V. (2021). Older Adults' Think-Aloud Verbalizations and Speech Features for Identifying User Experience Problems. In Association for Computing Machinery (Ed.), *CHI 2021: 358* (pp. 1–13). https://doi.org/10.1145/3411764.3445680

Ferrante, G., Licari, A., Marseglia, G. L., & La Grutta, S. (2021). Digital health interventions in children with asthma. *Clinical & Experimental Allergy*, 51(2), 212–220. https://doi.org/10.1111/cea.13793

Frøkjær, E., & Hornbæk, K. (2005). Cooperative usability testing: complementing usability tests with user-supported interpretation sessions. *Proceedings of the CHI Conference on Human Factors in Computing Systems, CHI 2005, Portland, OR, USA*, pp. 1383–1386. https://doi.org/10.1145/1056808.1056922

Fryer, K., Delgado, A., Foti, T., Reid, C. N., & Marshall, J. (2020). Implementation of Obstetric Telehealth During COVID-19 and Beyond. *Maternal and Child Health Journal*, 24, 1104–1110. https://doi.org/10.1007/s10995-020-02967-7

Fuchs, A., Stolze, C., & Thomas, O. (2013). Von der klassischen zur agilen Softwareentwicklung. Evolution der Methoden am Beispiel eines Anwendungssystems. *HMD Praxis der Wirtschaftsinformatik*, 50, 17–26. https://doi.org/10.1007/BF03340792

Garcia, A. C., & de Lara, S. M. A. (2018). Enabling Aid in Remote Care for Elderly People via Mobile Devices: The MobiCare Case Study. In Association for Computing Machinery (Ed.), *DSAI 2018: Proceedings of the 8th International Conference on Software Development and Technologies for Enhancing Accessibility and Fighting Info-exclusion* (pp. 270–277). https://doi.org/10.1145/3218585.3218671

Georgsson, M. (2020). A Review of Usability Methods Used in the Evaluation of Mobile Health Applications for Diabetes. *Studies in Health Technology and Informatics*, 273, 228–233.

Georgsson, M., Staggers, N., Årsand, E., & Kushniruk, A. (2019). Employing a user-centered cognitive walkthrough to evaluate a mHealth diabetes self-management application: A case study and beginning method validation. *Journal of Biomedical Informatics*, 91, 103110. https://doi.org/10.1016/j.jbi.2019.103110

Gilbert, B. J., Goodman, E., Chadda, A., Hatfield, D., Forman, D. E., & Panch, T. (2015). The Role of Mobile Health in Elderly Populations. *Current Geriatrics Reports*, 4, 347–352. https://doi.org/10.1007/s13670-015-0145-6

Góngora Alonso, S., de la Torre Díez, I., & Zapiraín, B. G. (2019). Predictive, Personalized, Preventive and Participatory (4P) Medicine Applied to Telemedicine and eHealth in the Literature. *Journal of Medical Systems*, 43(5), 140. https://doi.org/10.1007/s10916-019-1279-4

Gren, L., & Lenberg, P. (2020). Agility is responsiveness to change: An essential definition. In Association for Computing Machinery (Ed.), *International Conference Proceeding Series*

(ICPS). EASE 2020: Proceedings of the Evaluation and Assessment in Software Engineering (pp. 348–353). https://doi.org/10.1145/3383219.3383265

Gundelsweiler, F., Memmel, T., & Reiterer, H. (2004). Agile Usability Engineering. In R. Keil-Slawik, H. Selke, G. Szwillus (Eds.), *Mensch & Computer 2004: Allgegenwärtige Interaktion* (pp. 33–42). Oldenburg Verlag, München. https://doi.org/10.1524/9783486598773.33

Haggerty, T., Brabson, L., Grogg, K. A., Herschell, A. D., Giacobbi, P., Sedney C., & Dino, G. (2021). Usability testing of an electronic health application for patient activation on weight management. *mHealth*, 7(45), 1–9. https://doi.org/10.21037/mhealth-20-119

Haluza, D., Naszay, M., Stockinger, A., & Jungwirth, D. (2016). Prevailing Opinions on Connected Health in Austria: Results from an Online Survey. *International Journal of Environmental Research and Public Health*, 13(8), 813. https://doi.org/10.3390/ijerph13080813

Hanser, E. (2010). *Agile Prozesse: Von XP über Scrum bis MAP.* Springer. https://doi.org/10.1007/978-3-642-12313-9

Hassan, M. M., Tukiainen, M., & Qureshi, A. N. (2019). (Un)Discounted Usability: Evaluating Low-Budget Educational Technology Projects with Dual-Personae Evaluators. In Association for Computing Machinery (Ed.), *ICSIE 2019: Proceedings of the 8th International Conference on Software and Information Engineering* (pp. 253–258). https://doi.org/10.1145/3328833.3328860

Hedvall, P. O. (2009). Towards the Era of Mixed Reality: Accessibility Meets Three Waves of HCI. In A. Holzinger & K. Miesenberger (Eds.), *Lecture Notes in Computer Science: Vol. 5889. HCI and Usability for e-Inclusion* (pp. 264–278). Springer. https://doi.org/10.1007/978-3-642-10308-7_18

Heimgärtner, R. (2017). Usability-Engineering. In R. Heimgärtner (Ed.), *Interkulturelles User Interface Design. Von der Idee zum erfolgreichen Produkt* (pp. 81–133). Springer. https://doi.org/10.1007/978-3-662-48370-1

Helfferich, C. (2011). *Die Qualität qualitativer Daten. Manual für die Durchführung qualitativer Interviews.* VS Verlag für Sozialwissenschaften, Springer. https://doi.org/10.1007/978-3-531-92076-4

Hill, J. R., Harrington, A. B., Adeoye, P., Campbell, N. L., & Holden, R. J. (2021). Going Remote—Demonstration and Evaluation of Remote Technology Delivery and Usability Assessment With Older Adults: Survey Study. *JMIR mHealth and uHealth*, 9(3), e26702. https://doi.org/10.2196/26702

Holzinger, A., Popova, E., Peischl, B., & Ziefle, M. (2012). On Complexity Reduction of User Interfaces for Safety-Critical Systems. In G. Quirchmayr, J. Basl, I. You, L. Xu, & E. Weippl (Eds.), *Lecture Notes in Computer Science: Vol. 7465. Multidisciplinary Research and Practice for Information Systems* (pp. 108–121). Springer. https://doi.org/10.1007/978-3-642-32498-7_9

Hussain, Z., Slany, W., & Holzinger, A. (2009a). Current State of Agile User-Centered Design: A Survey. In A. Holzinger & K. Miesenberger (Eds.), *Lecture Notes in Computer Science: Vol. 5889. HCI and Usability for e-Inclusion* (pp. 416–427). Springer. https://doi.org/10.1007/978-3-642-10308-7_30

Hussain, Z., Slany, W., & Holzinger, A. (2009b). Investigating Agile User-Centered Design in Practice: A Grounded Theory Perspective. In A. Holzinger & K. Miesenberger (Eds.), *Lecture Notes in Computer Science: Vol. 5889. HCI and Usability for e-Inclusion* (pp. 279–289). Springer. https://doi.org/10.1007/978-3-642-10308-7_19

Institute of Medicine (2001). *Crossing the quality chasm: a new health system for the 21st century.* National Academy Press.

International Standardization Organization. (2018). *Ergonomics of human-system interaction—Part 11: Usability: Definitions and concepts (ISO 9241-11:2018(en)).* International Standardization Organization Geneva.

Iorfino, F., Cross, S. P., Davenport, T., Carpenter, J. S., Scott, E., Shiran, S., & Hickie, I. B. (2019). A Digital Platform Designed for Youth Mental Health Services to Deliver Personalized and Measurement-Based Care. *Frontiers in Psychiatry*, 10, 595. https://doi.org/10.3389/fpsyt.2019.00595

Isaković, M., Sedlar, U., Volk, M., & Bešter, J. (2016). Usability Pitfalls of Diabetes mHealth Apps for the Elderly. *Journal of Diabetes Research* 2016, 1–9. https://doi.org/10.1155/2016/1604609

Jakkaew, P., & Hongthong, T. (2017). Requirements Elicitation to Develop Mobile Application for Elderly. *11th Asia Pacific Conference, ICDAMT 2017, Chiang Mai, Thailand*, 464–467. https://doi.org/10.1109/ICDAMT.2017.7905013

Jaspers, M. W. M. (2009). A comparison of usability methods for testing interactive health technologies: Methodological aspects and empirical evidence. *International Journal of Medical Informatics*, 78(5), 340–353. https://doi.org/10.1016/j.ijmedinf.2008.10.002

Johnson C. M., Johnston D., Crowley P. K., Culbertson, H., Rippen, H. E., & Damico, D. J. (2011). *EHR Usability Toolkit: A Background Report on Usability and Electronic Health Records.* AHRQ Publication No. 11-0084-EF. Westat. https://digital.ahrq.gov/sites/default/files/docs/citation/EHR_Usability_Toolkit_Background_Report.pdf

Joyce, A. (2019, July 28). *Formative vs. Summative Evaluations.* Retrieved November 8, 2022, from https://www.nngroup.com/articles/formative-vs-summative-evaluations/

Kip, H., Keizer, J., da Silva, M. C., Beerlage-de Jong, N., Köhle, N., & Kelders, S. M. (2022). Methods for Human-Centered eHealth Development: Narrative Scoping Review. *Journal of Medical Internet Research*, 24(1), Article e31858. https://doi.org/10.2196/31858

Knapp, C., Madden, V., Wang, H., Sloyer, P., & Shenkman, E. (2011). Internet use and eHealth literacy of low-income parents whose children have special health care needs. *Journal of Medical Internet Research*, 13(3), Article e75. https://doi.org/10.2196/jmir.1697

Kuckartz, U. (2014). *Mixed Methods. Methodologie, Forschungsdesigns und Analyseverfahren.* Springer Fachmedien. https://doi.org/10.1007/978-3-531-93267-5

Kushniruk, A., & Borycki, E. (2006). Low-cost rapid usability engineering: designing and customizing usable healthcare information systems. *Healthcare quarterly*, 9(4), 98–102.

Kushniruk, A. W., & Borycki, E. M. (2015). Integrating Low-Cost Rapid Usability Testing into Agile System Development of Healthcare IT: A Methodological Perspective. *Studies in Health Technology and Informatics*, 210, 200–4.

Kushniruk, A., & Borycki, E. (2017). Low-Cost Rapid Usability Testing: Its Application in Both Product Development and System Implementation. *Studies in Health Technology Informatics*, 234, 195–200.

Kushniruk, A. W., & Patel, V. L. (2004). Cognitive and usability engineering methods for the evaluation of clinical information systems. *Journal of Biomedical Informatics*, 37(1), 56–76. https://doi.org/10.1016/j.jbi.2004.01.003

Kuziemsky, C. E., & Kushniruk, A. (2014). Context Mediated Usability Testing. *Studies in Health Technology and Informatics*, 205, 905–9.

Lau, F., & Kuziemsky, C. (2016). *Handbook of eHealth Evaluation: An Evidence-based Approach*. University of Victoria.

Lazard, A. J., Saffer, A. J., Horrell, L., Benedict, C., & Love, B. (2019). Peer-to-peer connections: Perceptions of a social support app designed for young adults with cancer. *Psycho-Oncology*, 29(1), 173–181. https://doi.org/10.1002/pon.5220

Leahy, D., & Dolan, D. (2009). Digital Literacy—Is It Necessary for eInclusion? In A. Holzinger & K. Miesenberger (Eds.), *Lecture Notes in Computer Science: Vol. 5889. HCI and Usability for e-Inclusion* (pp. 149–158). Springer. https://doi.org/10.1007/978-3-642-10308-7_10

Lewis, J. R. (2014). Usability: Lessons Learned ... and Yet to Be Learned. *International Journal of Human-Computer Interaction*, 30(9), 663–684. https://doi.org/10.1080/10447318.2014.930311

Liveri, D., Sarri, A., & Skouloudi, C. (Eds.). (2015). *Security and Resilience in eHealth. Security Challenges and Risks*. Iraklio, Greece: European Union Agency for Network and Information Security (ENISA). https://doi.org/10.2824/217830

Lum, A. S. L., Chiew, T. K., Ng, C. J., Lee, Y. K., Lee, P. Y., & Teo, C. H. (2017). Development of a web-based insulin decision aid for the elderly: usability barriers and guidelines. *Universal Access in the Information Society*, 16(3), 775–791. https://doi.org/10.1007/s10209-016-0503-y

Magües, D. A., Castro, J. W., & Acuña, S. T. (2016). HCI usability techniques in agile development. *Proceedings of the IEEE International Conference on Automatica, ICA-ACCA, Curicó, Chile*, 1–7. https://doi.org/10.1109/ICA-ACCA.2016.7778513

Maramba, I., Chatterjee, A., & Newman, C. (2019). Methods of usability testing in the development of eHealth applications: A scoping review. *International Journal of Medical Informatics*, 126, 95–104. https://doi.org/10.1016/j.ijmedinf.2019.03.018

Marcilly, R., Ammenwerth, E., Roehrer, E., Pelayo, S., Vasseur, F., & Beuscart-Zéphir, M.-C. (2015). Usability Flaws in Medication Alerting Systems: Impact on Usage and Work System. *Yearbook of Medical Informatics*, 24(1), 55–67. https://doi.org/10.15265/IY-2015-006

Marcilly, R., Schiro, J., Heyndels, L., Guerlinger, S., Pigot, A., & Pelayo, S. (2021). Competitive Usability Evaluation of Electronic Health Records: Preliminary Results of a Case Study. *Studies in Health Technology and Informatics*, 834–8. https://doi.org/10.3233/SHTI210296

Marcus, A. (2003). Universal, Ubiquitous, User-Interface Design for the Disabled and Elderly. *Interactions*, 10(2), 23–7. https://doi.org/10.1145/637848.637858

Mayberry, L. S., Lyles, C. R., Oldenburg, B., Osborn, C. Y., Parks, M., & Peek, M. E. (2019). mHealth Interventions for Disadvantaged and Vulnerable People with Type 2 Diabetes. *Current Diabetes Reports*, 19(12), 148. https://doi.org/10.1007/s11892-019-1280-9

Mayring, P. (2015). *Qualitative Inhaltsanalyse. Grundlagen und Techniken*. Beltz Verlag.

McCann, L., McMillan, K. A., & Pugh, G. (2019). Digital Interventions to Support Adolescents and Young Adults With Cancer: Systematic Review. *JMIR Cancer*, 5(2), Article e12071. https://doi.org/10.2196/12071

Momenipour, A., Rojas-Murillo, S., Murphy, B., Pennathur, P., & Pennathur, A. (2021). Usability of state public health department websites for communication during a pandemic: A heuristic evaluation. *International Journal of Industrial Ergonomics*, 86, 103216. https://doi.org/10.1016/j.ergon.2021.103216

Moran, K. (2019, December 1). *Usability Testing 101.* Retrieved November 9, 2022, from https://www.nngroup.com/articles/usability-testing-101/

Moreno, L., Martínez, P., & Ruiz-Mezcua, B. (2009). A Bridge to Web Accessibility from the Usability Heuristics. In A. Holzinger & K. Miesenberger (Eds.), *Lecture Notes in Computer Science: Vol. 5889. HCI and Usability for e-Inclusion* (pp. 290–300). Springer. https://doi.org/10.1007/978-3-642-10308-7_20

Mori, G., Buzzi, M. C., Buzzi, M., Leporini, B., & Penichet, V. M. R. (2011). Collaborative Editing for All: The Google Docs Example. In C. Stephanidis (Ed.) *Lecture Notes in Computer Science: Vol. 6768. Universal Access in Human-Computer Interaction. Applications and Services* (pp. 165–174). Springer. https://doi.org/10.1007/978-3-642-21657-2_18

Möller, S. (2017). *Quality Engineering. Qualität kommunikationstechnischer Systeme.* Springer. https://doi.org/10.1007/978-3-662-56046-4

Munn, Z., Peters, M. D. J., Stern, C., Tufanaru, C., McArthur, A., & Aromataris, E. (2018a). Systematic review or scoping review? Guidance for authors when choosing between a systematic or scoping review approach. *BMC Medical Research Methodology*, 18, 143. https://doi.org/10.1186/s12874-018-0611-x

Munn, Z., Stern, C., Aromataris, E., Lockwood, C., & Jordan, Z. (2018b). What kind of systematic review should I conduct? A proposed typology and guidance for systematic reviewers in the medical and health sciences. *BMC Medical Research Methodology*, 18(1), 5. https://doi.org/10.1186/s12874-017-0468-4

Nielsen, J. (1992). Finding usability problems through heuristic evaluation. In P. Bauersfeld, J. Bennett, & G. Lynch (Eds.), *CHI 1992: Proceedings of the SIGCHI Conference on Human Factors in Computing Systems* (pp. 373–380). Association for Computing Machinery. https://doi.org/10.1145/142750.142834

Nielsen, J. (1994). Usability inspection methods. In C. Plaisant (Ed.), *CHI 1994: Proceedings of the Conference Companion on Human Factors in Computing Systems* (pp. 413–414). Association for Computing Machinery. https://doi.org/10.1145/259963.260531

Nunes, F., Andersen, T., & Fitzpatrick, G. (2019). The agency of patients and carers in medical care and self-care technologies for interacting with doctors. *Health Informatics Journal*, 25(2), 330–349. https://doi.org/10.1177/1460458217712054

Nunes, F., Kerwin, M., & Silva, P. A. (2012). Design recommendations for tv user interfaces for older adults: findings from the eCAALYX project. In Association for Computing Machinery (Ed.), *International Conference on Computers and accessibility. ASSETS 2012: Proceedings of the 14th international ACM SIGACCESS conference on Computers and accessibility* (pp. 41–8). https://doi.org/10.1145/2384916.2384924

Obendorf, H., Schmolitzky, A., & Finck, M. (2018). *XPnUE—Defining and Teaching a Fusion of eXtreme Programming & Usability Engineering.*

Oikonomou, T., Votis, K., Korn, P., Tzovaras, D., & Likothanasis, S. (2009). A Standalone Vision Impairments Simulator for Java Swing Applications. In A. Holzinger & K. Miesenberger (Eds.), *Lecture Notes in Computer Science: Vol. 5889. HCI and Usability for e-Inclusion* (pp. 387–398). Springer. https://doi.org/10.1007/978-3-642-10308-7_27

Orimo, H., Ito, H., Suzuki, T., Araki, A., Hosoi, T., & Sawabe, M. (2006). Reviewing the definition of "elderly". *Geriatrics & Gerontology International*, 6(3), 149–158. https://doi.org/10.1111/j.1447-0594.2006.00341.x

Page, M. J., McKenzie, J. E., Bossuyt, P. M., Boutron, I., Hoffmann, T. C., Mulrow, C. D., Shamseer, L., Tetzlaff, J. M., Akl, E. A., Brennan, S. E., Chou, R., Glanville, J., Grimshaw,

J. M., Hróbjartsson, A., Lalu, M. M., Li, T., Loder, E. W., Mayo-Wilson, E., McDonald, S., McGuinness, L. A., Stewart, L. A., Thomas, J., Tricco, A. C., Welch, V. A., Whiting, P., & Moher, D. (2021). The PRISMA 2020 statement: an updated guideline for reporting systematic reviews. *The British Medical Journal*, 372, Article n71. https://doi.org/10.1136/bmj.n71

Parry, D., Carter, P., Koziol-McLain, J., & Feather, J. (2015). A Model for Usability Evaluation for the Development and Implementation of Consumer eHealth Interventions. *Studies in Health Technology and Informatics*, 2016, 968.

Pawson, M., & Greenberg, S. (2009). Extremely rapid usability testing. *Journal of Usability Studies*, 4(3), 124–135.

Peute, L. W., Knijnenburg, S. L., Kremer, L. C., & Jaspers, M. W. M. (2015). A Concise and Practical Framework for the Development and Usability Evaluation of Patient Information Websites. *Applied Clinical Informatics*, 6(2), 383–399. https://doi.org/10.4338/ACI-2014-11-RA-0109

Pohl, M., Endl, H., & Fels, U. (2016). Animated Scatterplot—Analysis of Time-Oriented Data of Diabetes Patients. *Studies in Health Technology and Informatics*, 223, 191–8. https://doi.org/10.3233/978-1-61499-645-3-191

Pohl, M., & Scholz, F. (2014). How to Investigate Interaction with Information Visualisation: An Overview of Methodologies. In A. Ebert, G. van der Veer, G. Domik, N. Gershon, & I. Scheler (Eds.) *Lecture Notes in Computer Science: Vol. 8345. Building Bridges: HCI, Visualization, and Non-formal Modeling* (pp. 17–29). Springer. https://doi.org/10.1007/978-3-642-54894-9_3

Price, M., Weber, J. H., Davies, I., & Bellwood, P. (2015). Lead User Design: Medication Management in Electronic Medical Records. In I. N. Sarkar, A. Georgiou & P. Mazzoncini de Azevedo Marques (Eds.) *Studies in Health Technology and Informatics* (pp. 237–241). https://doi.org/10.3233/978-1-61499-564-7-237

Rafi, U., Mustafa, T. M., Iqbal, N., & Zafar, W. (2015). US-Scrum: A Methodology for Developing Software with Enhanced Correctness, Usability and Security. *International Journal of Scientific and Engineering Research*, 6(9), 377–383.

Ramsey, W. A., Heidelberg, R. E., Gilbert, A. M., Heneghan, M. B., Badawy, S. M., & Alberts, N. M. (2020). eHealth and mHealth interventions in pediatric cancer: A systematic review of interventions across the cancer continuum. *Psycho-Oncology*, 29(1), 17–37. https://doi.org/10.1002/pon.5280

Razak, F. H. A., Razak, N. A., Adnan, W. A. W., & Ahmad, N. A. (2013). How Simple is Simple: Our Experience with Older Adult Users. In Association for Computing Machinery (Ed.), *International Conference Proceeding Series (ICPS). APCHI 2013: Proceedings of the 11th Asia Pacific Conference on Computer Human Interaction* (pp. 379–387). https://doi.org/10.1145/2525194.2525307

Richter, M., & Flückinger, M. D. (2016). *Usability und UX kompakt. Produkte für Menschen.* Springer-Verlag. https://doi.org/10.1007/978-3-662-49828-6

Rietze, J., Bürkner, I., Pfister, A., & Blum, R. (2021). Online Focus Groups with and for the Elderly: Specifics, Challenges, Recommendations: Online-Fokusgruppen mit und für Senior*innen: Besonderheiten, Herausforderungen, Empfehlungen. In Association for Computing Machinery (Ed.), *MuC 2021: Proceedings of Mensch und Computer 2021* (pp. 194–198). https://doi.org/10.1145/3473856.3474300

Roberts, V. L., & Fels, D. I. (2006). Methods for inclusion: Employing think aloud protocols in software usability studies with individuals who are deaf. *International Journal of Human-Computer Studies*, 64(6), 489–501. https://doi.org/10.1016/j.ijhcs.2005.11.001

Rogers, Y. (2009). The Changing Face of Human-Computer Interaction in the Age of Ubiquitous Computing. In A. Holzinger & K. Miesenberger (Eds.), *Lecture Notes in Computer Science: Vol. 5889. HCI and Usability for e-Inclusion* (pp. 1–19). Springer. https://doi.org/10.1007/978-3-642-10308-7_1

Rödle, W., Wimmer, S., Zahn, J., Prokosch, H., Hinkes, B., Neubert, A., Rascher, W., Kraus, S., Toddenroth, D., & Sedlmayr, B. (2019). User-Centered Development of an Online Platform for Drug Dosing Recommendations in Pediatrics. *Applied Clinical Informatics*, 10(4), 570–9. https://doi.org/10.1055/s-0039-1693714

Russ, A. L., Baker, D. A., Fahner, W. J., Milligan, B. S., Cox, L., Hagg, H. K., & Saleem, J. J. (2010). A Rapid Usability Evaluation (RUE) Method for Health Information Technology. *Proceedings of the American Medical Informatics Association Symposium, AMIA 2010, Washington, DC, USA*, 702–6.

Schoeberlein, J. G., & Wang, Y. K. (2011). Examining the Current State of Group Support Accessibility: A Focus Group Study. In C. Stephanidis (Ed.) *Lecture Notes in Computer Science: Vol. 6768. Universal Access in Human-Computer Interaction. Applications and Services* (pp. 272–281). Springer. https://doi.org/10.1007/978-3-642-21657-2_29

Schoor, B. (2022, January 5). *Agile vs. agility. What are the differences?* AGILEXL. Retrieved November 9, 2022, from https://agilexl.com/blog/agile-vs-agility-what-are-the-difference

Schreier, M., & Odağ, Ö. (2010). Mixed Methods. In G. Mey, & K. Mruck (Eds.), *Handbuch Qualitative Forschung in der Psychologie* (pp. 263–277). VS Verlag für Sozialwissenschaften, Springer. https://doi.org/10.1007/978-3-531-92052-8_18

Schwaninger, I., Carros, F., Weiss, A., Wulf, V., & Fitzpatrick, G. (2022). Video connecting families and social robots: from ideas to practices putting technology to work. *Universal Access in the Information Society*, 22, 931–943. https://doi.org/10.1007/s10209-022-00901-y

Setia, M. S. (2016). Methodology Series Module 5: Sampling Strategies. *Indian Journal of Dermatology*, 61(5), 505–9.

Silva, A., Martins, A. I., Caravau, H., Almeida, A. M., Silva, T., Ribeiro, Ó., Santinha, G., & Rocha, N. (2020). Experts Evaluation of Usability for Digital Solutions Directed at Older Adults: a Scoping Review of Reviews. In Association for Computing Machinery (Ed.), *International Conference Proceeding Series (ICPS). DSAI 2020: Proceedings of 9th International Conference on Software Development and Technologies for Enhancing Accessibility and Fighting Info-exclusion* (pp. 174–181). https://doi.org/10.1145/3439231.3439238

Sinabell, I., & Ammenwerth, E. (2022). ToUsE: Toolbox for eHealth Usability Evaluations. Austria, Hall in Tirol: UMIT TIROL. *Available in the ESM.*

Singh, M. (2008). U-SCRUM: An Agile Methodology for Promoting Usability. *Proceedings of the Agile 2008 Development Conference, Toronto, ON, Canada*, 555–560. https://doi.org/10.1109/Agile.2008.33

Soares, M. M., Rosenzweig, E., & Marcus, A. (Eds.). (2022). *Lecture Notes in Computer Science: Vol. 13322. Design, User Experience, and Usability: Design for Emotion, Well-being and Health, Learning, and Culture.* Springer. https://doi.org/10.1007/978-3-031-05906-3

Sohaib, O., & Khan, K. (2010). Integrating usability engineering and agile software development: A literature review. *Proceedings of the International Conference on Computer Design and Applications, ICCDA 2010, Qinhuangdao, China*, 32–8. https://doi.org/10.1109/ICCDA.2010.5540916

Sounderajah, V., Clarke, J., Yalamanchili, S., Acharya, A., Markar, S. R., Ashrafian, H., & Darzi, A. (2021). A national survey assessing public readiness for digital health strategies against COVID-19 within the United Kingdom. *Scientific Reports*, 11, 5958. https://doi.org/10.1038/s41598-021-85514-w

Spanakis, E. G., Santana, S., Tsiknakis, M., Marias, K., Sakkalis, V., Teixeira, A., Janssen, J. H., de Jong, H., & Tziraki, C. (2016). Technology-Based Innovations to Foster Personalized Healthy Lifestyles and Well-Being: A Targeted Review. *Journal of Medical Internet Research*, 18(6), Article e128. https://doi.org/10.2196/jmir.4863

Spiel, K., Brulé, E., Frauenberger, C., Bailley, G., & Fitzpatrick, G. (2020). In the details: the micro-ethics of negotiations and in-situ judgements in participatory design with marginalised children. *CoDesign*, 16(1), 45–65. https://doi.org/10.1080/15710882.2020.1722174

Spiliotopoulos, D., Stavropoulou, P., & Kouroupetroglou, G. (2009). Spoken Dialogue Interfaces: Integrating Usability. In A. Holzinger & K. Miesenberger (Eds.), *Lecture Notes in Computer Science: Vol. 5889. HCI and Usability for e-Inclusion* (pp. 484–499). Springer. https://doi.org/10.1007/978-3-642-10308-7_36

Springett, M., Mihajlov, M., Brzovska, E., Orozel, M., Elsner, V., Oppl, S., Stary, C., Keith, S., & Richardson, J. (2021). An analysis of social interaction between novice older adults when learning gesture-based skills through simple digital games. *Universal Access in the Information Society*, 21, 639–655. https://doi.org/10.1007/s10209-021-00793-4

Stapelkamp, T. (2010). *Interaction- und Interfacedesign. Web-, Game-, Produkt- und Servicedesign Usability und Interface als Corporate Identity.* Springer-Verlag Berlin Heidelberg. https://doi.org/10.1007/978-3-642-02074-2

Stara, V., Vera, B., Bollinger, D., Rossi, L., Felici, E., Di Rosa, M., de Jong, M., & Paolini, S. (2021). Usability and Acceptance of the Embodied Conversational Agent Anne by People With Dementia and Their Caregivers: Exploratory Study in Home Environment Settings. *JMIR mHealth and uHealth*, 9(6), Article e25891. https://doi.org/10.2196/25891

Stephanidis, C., & Akoumianakis, D. (2001). Universal design: Towards universal access in the information society. In Association for Computer Machinery (Ed.), *CHI EA 2001: CHI 2001 Extended Abstracts on Human Factors in Computing Systems* (pp. 499–500). https://doi.org/10.1145/634067.634352

Sülz, S., van Elten, H. J., Askari, M., Weggelaar-Jansen, A. M., & Huijsman, R. (2021). eHealth Applications to Support Independent Living of Older Persons: Scoping Review of Costs and Benefits Identified in Economic Evaluations. *Journal of Medical Internet Research*, 23(3), Article e24363. https://doi.org/10.2196/24363

The User Experience Professionals' Association (UXPA). (2010). *Usability Body of Knowledge.* Retrieved November 9, 2022, from http://www.usabilitybok.org/methods

Thielsch, M. T., & Salaschek, M. (2020). Toolbox zur Website-Evaluation: Erfassung der User Experience von Onlinegesundheitsinformationen. *Bundesgesundheitsblatt—Gesundheitsforschung—Gesundheitsschutz*, 63, 721–8. https://doi.org/10.1007/s00103-020-03142-7

Tsonev, N. (2021, September 21). *Agile vs Agility: How to Be Truly Adaptable to Changes?* Kanbanize. Retrieved November 9, 2022, from https://kanbanize.com/blog/agile-vs-agility/

University of Minnesota Duluth (UMD). (2022). *Usability Evaluation Toolbox.* Retrieved October 31, 2022, from https://www.d.umn.edu/itss/training/online/usability/toolbox.html

Utesheva, A., & Boell, S. K. (2016). Theorizing Society and Technology in Information Systems Research. In Association for Computer Machinery (Ed.), *ACM SIGMIS Database: the DATABASE for Advances in Information Systems*, 47(4), (pp. 106–110). https://doi.org/10.1145/3025099.3025110

van den Heuvel, J. FM., Groenhof, T. K., Veerbeek, J. HW., van Solinge, W. W., Lely, A. T., Franx, A., & Bekker, M. N. (2018). eHealth as the Next-Generation Perinatal Care: An Overview of the Literature. *Journal of Medical Internet Research*, 20(6), Article e202. https://doi.org/10.2196/jmir.9262

Vanderheiden, G. (2000). Fundamental principles and priority setting for universal usability. In J. Thomas (Ed.), *CUU 2000: Proceedings on the 2000 conference on Universal Usability* (pp. 32–7). Association for Computing Machinery. https://doi.org/10.1145/355460.355469

Verhoeven, F., & van Gemert-Pijnen, J. (2010). Discount User-Centered e-Health Design: A Quick-but-not-Dirty Method. In G. Leitner, M. Hitz, & A. Holzinger (Eds.), *Lecture Notes in Computer Science: Vol. 6389. HCI in Work and Learning, Life and Leisure* (pp. 101–123). Springer. https://doi.org/10.1007/978-3-642-16607-5_7

Villani, D., Cognetta, C., Toniolo, D., Scanzi, F., & Riva, G. (2016). Engaging Elderly Breast Cancer Patients: The Potential of eHealth Interventions. *Frontiers in Psychology*, 7, 1825. https://doi.org/10.3389/fpsyg.2016.01825

Vitiello, G., & Sebillo, M. (2018). The importance of empowerment goals in elderly-centered interaction design. *Proceedings of the International Conference on Advanced Visual Interfaces, AVI 2018, Castiglione della Pescaia, Grosseto, Italy*, 37, 1–5. https://doi.org/10.1145/3206505.3206551

Wang, Q., Zhou, L., Liu, J., Tian, J., Chen, X., Zhang, W., Wang, H., Zhou, W., & Gao, Y. (2022). Usability evaluation of mHealth Apps for the elderly: A scoping review. *BMC Medical Informatics and Decision Making* (preprint). https://doi.org/10.21203/rs.3.rs-1515149/v1

Wernhart, A., Gahbauer, S., & Haluza, D. (2019). eHealth and telemedicine: Practices and beliefs among healthcare professionals and medical students at a medical university. *PLOS ONE*, 14(2), e0213067. https://doi.org/10.1371/journal.pone.0213067

Wildenbos, G. A. (2019a). *Design speaks: Improving patient-centeredness for older people in a digitalizing healthcare context.* [Doctoral dissertation, University of Amsterdam].

Wildenbos, G. A., Jaspers, M. W. M., Schijven, M. P., & Dusseljee-Peute, L. W. (2019b). Mobile health for older adult patients: Using an aging barriers framework to classify usability problems. *International Journal of Medical Informatics*, 124, 68–77. https://doi.org/10.1016/j.ijmedinf.2019.01.006

Wildenbos, G. A., Peute, L., & Jaspers, M. (2018). Aging barriers influencing mobile health usability for older adults: A literature based framework (MOLD-US). *International Journal of Medical Informatics*, 114, 66–75. https://doi.org/10.1016/j.ijmedinf.2018.03.012

Wozney, L. M., Baxter, P., Fast, H., Cleghorn, L., Hundert, A. S., & Newton, A. S. (2016). Sociotechnical Human Factors Involved in Remote Online Usability Testing of Two eHealth Interventions. *JMIR Human Factors*, 3(1), Article e6.

Wronikowska, M. W., Malycha, J., Morgan, L. J., Westgate, V., Petrinic, T., Young, J. D., & Watkinson, P. J. (2021). Systematic review of applied usability metrics within usability evaluation methods for hospital electronic healthcare record systems. *Journal of Evaluation in Clinical Practice*, 27(6), 1403–16. https://doi.org/10.1111/jep.13582

Yaddaden, A., Bier, N., Andrée-Anne, P., Lussier, M., Aboujaoudé, A., & Gagnon-Roy, M. (2019). Usability questionnaires for telemonitoring and assistive technology for cognition for older adults: a rapid review. In Association for Computing Machinery (Ed.), *GoodTechs 2019: Proceedings of the 5th EAI International Conference on Smart Objects and Technologies for Social Good* (pp. 102–7). https://doi.org/10.1145/3342428.3342653

Yin, R. K. (2003). *Case Study Research. Design and Methods.* SAGE Publications, California, USA.

Printed in the United States
by Baker & Taylor Publisher Services